Human Dynamics in Smart Cities

Series editors

Shih-Lung Shaw, Department of Geography, University of Tennessee, Knoxville, TN, USA
Daniel Sui, Department of Geography, Ohio State University, Columbus, OH, USA

This series covers advances in information and communication technology (ICT), mobile technology, and location-aware technology and ways in which they have fundamentally changed how social, political, economic and transportation systems work in today's globally connected world. These changes have raised many exciting research questions related to human dynamics at both disaggregate and aggregate levels that have attracted attentions of researchers from a wide range of disciplines. This book series aims to capture this emerging dynamic interdisciplinary field of research as a one-stop depository of our cumulative knowledge on this topic that will have profound implications for future human life in general and urban life in particular. This book series plans to cover topics from theoretical perspectives, space-time analytics, modeling human dynamics, urban analytics, social media and big data, travel dynamics, to privacy issues, development of smart cities, and problems and prospects of human dynamics research. We invite contributions of theoretical, technical, or application aspects of human dynamics research for a global and interdisciplinary audience.

More information about this series at http://www.springer.com/series/15897

Shih-Lung Shaw · Daniel Sui
Editors

Human Dynamics Research in Smart and Connected Communities

 Springer

Editors
Shih-Lung Shaw
Department of Geography
University of Tennessee
Knoxville, TN
USA

Daniel Sui
Department of Geography
Ohio State University
Columbus, OH
USA

ISSN 2523-7780   ISSN 2523-7799   (electronic)
Human Dynamics in Smart Cities
ISBN 978-3-319-73246-6   ISBN 978-3-319-73247-3   (eBook)
https://doi.org/10.1007/978-3-319-73247-3

Library of Congress Control Number: 2018930114

Printed on acid-free paper

This Springer imprint is published by the registered company Springer International Publishing AG part of Springer Nature
The registered company address is: Gewerbestrasse 11, 6330 Cham, Switzerland

Contents

Editors and Contributors

About the Editors

Shih-Lung Shaw is Alvin and Sally Beaman Professor and Arts and Sciences Excellence Professor of Geography at the University of Tennessee, Knoxville. He also serves as the Interim Associate Provost for international education at the University of Tennessee, Knoxville. He received his B.S. degree from the National Taiwan University and his M.A. and Ph.D. degrees from the Ohio State University. His research interests cover geographic information science (GIScience), transportation geography, time geography, GIS for transportation (GIS-T), and space–time analytics of human dynamics. His recent research has focused on space–time analytics of human activities and interactions in a hybrid physical–virtual world based on various types of individual tracking data such as cell phone data, online social media data, vehicle tracking data, travel–activity survey data, and population migration data. His research has led to the development of a space–time GIS for representation, analysis, and visualization of individual activities and interactions in a hybrid physical–virtual space. Dr. Shaw is a Fellow of the American Association for Advancement of Science (AAAS). He also received the Edward L. Ullman Award for outstanding contributions to transportation geography from the American Association of Geographers (AAG) and served as the Head of the Department of Geography at the University of Tennessee, Knoxville.

Daniel Sui is Arts and Sciences Distinguished Professor and Professor of Geography, Public Affairs, Public Health, and Urban/Regional Planning at the Ohio State University (OSU). Since July 2016, he has been on an IPA assignment to serve as the Division Director for Social and Economic Sciences (SES) at the US National Science Foundation. Prior his current appointment, Daniel Sui served as Chair of Geography (2011–2015) and as Director of the Center for Urban & Regional Analysis (CURA) (2009–2012) at OSU. He holds B.S. (1986) and M.S. (1989) degrees from Peking University and Ph.D. degree from University of Georgia (1993). His current research interests cover issues related to robust and reliable practices in spatially integrated social sciences and humanities, human dynamics in smart & connected cities, and location-based social media. Daniel Sui was a 2009 Guggenheim Fellow, 2014 Google Faculty Fellow, and 2014 recipient of the distinguished scholar award from the American Association of Geographers. Sui was also the 2015 Public Policy Scholar in residence at the Woodrow Wilson International Center for Scholars. More information about Sui's current research can be found at: https://www.researchgate.net/profile/Daniel_Sui.

Contributors

Budhendra Bhaduri Oak Ridge National Laboratory, Oak Ridge, TN, USA

Hanzhou Chen Department of Geography, Pennsylvania State University, State College, PA, USA

Jie Chen State Key Laboratory of Resources and Environmental Information System, Institute of Geographic Sciences and Natural Resources Research, Chinese Academy of Sciences, Beijing, People's Republic of China

Zhuo Chen Department of Geography, Kent State University, Kent, USA

Tao Cheng SpaceTimeLab for Big Data Analytics, Department of Civil, Environmental & Geomatic Engineering, University College London, London, UK

Lanxue Dang Department of Computer Science, Henan University, Kaifeng, China

Junchuan Fan Department of Geographical Sciences, University of Maryland, College Park, MD, USA

Song Gao Department of Geography, University of Wisconsin, Madison, WI, USA

Brendan Hoover Department of Geography and the Environment, The University of Texas at Austin, Austin, TX, USA

Ching-Shun Hsu National Taiwan University, Taipei, Taiwan

Yingjie Hu Department of Geography, University of Tennessee, Knoxville, TN, USA

Cheng-Chia Huang Department of Geography, San Diego State University, San Diego, CA, USA

Joe-Air Jiang National Taiwan University, Taipei, Taiwan

Jehn-Yih Juang National Taiwan University, Taipei, Taiwan

Caglar Koylu Department of Geographical and Sustainability Sciences, University of Iowa, Iowa City, IA, USA

Jay Lee Department of Geography, Kent State University, Kent, USA

Qingquan Li Shenzhen Key Laboratory of Spatial Smart Sensing and Services, Shenzhen University, Shenzhen, People's Republic of China

Feng Lu State Key Laboratory of Resources and Environmental Information System, Institute of Geographic Sciences and Natural Resources Research, Chinese Academy of Sciences, Beijing, People's Republic of China

Wei Luo School of Geographical Sciences and Urban Planning, Arizona State University, Tempe, AZ, USA

Huina Mao Oak Ridge National Laboratory, Oak Ridge, TN, USA

Jennifer A. Miller Department of Geography and the Environment, The University of Texas at Austin, Austin, TX, USA

Atsushi Nara Department of Geography, San Diego State University, San Diego, CA, USA

Jesse Piburn Oak Ridge National Laboratory, Oak Ridge, TN, USA

Shih-Lung Shaw Department of Geography, University of Tennessee, Knoxville, TN, USA

Jianan Shen SpaceTimeLab for Big Data Analytics, Department of Civil, Environmental & Geomatic Engineering, University College London, London, UK

Kelly Sims Oak Ridge National Laboratory, Oak Ridge, TN, USA

Kevin Sparks Oak Ridge National Laboratory, Oak Ridge, TN, USA

Kathleen Stewart Department of Geographical Sciences, University of Maryland, College Park, MD, USA

Robert Stewart Oak Ridge National Laboratory, Oak Ridge, TN, USA

Daniel Sui Department of Geography, Ohio State University, Columbus, OH, USA

Chih-Hong Sun National Taiwan University, Taipei, Taiwan

Gautam Thakur Oak Ridge National Laboratory, Oak Ridge, TN, USA

Ming-Hsiang Tsou Department of Geography, San Diego State University, San Diego, CA, USA

Marie Urban Oak Ridge National Laboratory, Oak Ridge, TN, USA

Eric Weber Oak Ridge National Laboratory, Oak Ridge, TN, USA

Tzai-Hung Wen National Taiwan University, Taipei, Taiwan

Chen Xu Department of Geography, University of Wyoming, Laramie, WY, USA

Yang Xu Department of Land Surveying and Geo-Informatics, The Hong Kong Polytechnic University, Kowloon, Hong Kong

Jiue-An Yang Qualcomm Institute, University of California, San Diego, CA, USA

Xinyue Ye Department of Geography, Kent State University, Kent, USA

Chapter 1
Introduction: Human Dynamics in Perspective

Shih-Lung Shaw and Daniel Sui

1.1 Introduction

Human dynamics have been in existence as long as the human history. The topic has been studied by researchers in many different disciplines over time. One characteristic of human dynamics is that they evolve with the changing environments, technologies, and human societies. Human dynamics in today's world are very different in many aspects from how we carried out our activities and interactions even a few decades ago. There are many factors contributing to the changing human dynamics. During the last three decades, advances in modern technologies such as information and communication (ICT) technology, location-aware technology, sensor technology, and mobile technology have played an important role in changing human activity and interaction patterns. For example, smartphones have enabled us to stay connected and interact with other people through a wide range of services and information available on the Internet to carry out various activities. Although the basic human needs have stayed the same as they were before, the ways we fulfill these activities have changed significantly due to modern technologies. It is now feasible to complete many office tasks from anywhere with an Internet connection and an appropriate device. When we are looking for a dinner place, we can quickly find a restaurant recommended by people we do not know via an app on our smartphone. We can walk to a bus stop just in time to catch a ride since the real-time bus locations are available at our fingertip. There is no doubt these changing human behaviors are leading to different kinds of dynamics in the urban, economic, transportation, social, and cultural systems.

S.-L. Shaw (✉)
Department of Geography, University of Tennessee, Knoxville, TN, USA
e-mail: sshaw@utk.edu

D. Sui
Department of Geography, The Ohio State University, Columbus, OH, USA

© Springer International Publishing AG, part of Springer Nature 2018
S.-L. Shaw and D. Sui (eds.), *Human Dynamics Research in Smart and Connected Communities*, Human Dynamics in Smart Cities,
https://doi.org/10.1007/978-3-319-73247-3_1

However, our knowledge about the implications of these changing human dynamics to our communities and societies is still limited.

Modern technologies not only have introduced changes to human dynamics but also have enabled our capability of collecting detailed data about human dynamics. Facebook knows who our friends are, how we interact with our friends, along with the timeline of our activities. Amazon keeps track of what we buy, how frequently we buy different items, and even the items we browsed. Google can use the keywords in our Gmail messages to help with targeted advertising. We also use Twitter, YouTube, Instagram, Foursquare, OpenStreetMap and many other apps to contribute information voluntarily. They are generally known as Volunteered Geographic Information (VGI) or crowdsourced data (Sui et al. 2013). Mobile phone service providers know where we have been, when we were there, whom we have contacted, and how frequently we contact various people through our communication records. Our bank knows what we buy, where and when we buy them, and how much we pay for each item through our credit card transactions. Although we normally do not intend to publicize our activities, data are nonetheless collected, analyzed, and even shared publically. Even we do not use a smartphone or any of the modern ICT devices, our data still can be collected via modern technologies. For example, an increasing number of security cameras mounted in public spaces can take pictures of passing people and vehicles. Through image processing plus facial recognition and license plate number matching software, people and vehicles can be identified and tracked. With cameras mounted on drones, privacy in our own fenced backyard also is in danger. Furthermore, our friends can post pictures of us on Facebook or Instagram even we do not use those apps. It therefore would be naïve to assume that we still have the same level of control of our privacy in today's world as we were a few decades ago. The reality we face today is a matter of who have our data and how they are using our data! In other words, it is out of our control to a great extent. This is part of the life we have to deal with in the Big Data era (Manyika et al. 2011). One challenge to human dynamics research community is how we can use the unprecedented data collection via various sources to help us gain insights on human dynamics in order to answer important questions to our communities and societies and make smart decisions for a better future of our communities and societies.

Human dynamics research faces many challenges of addressing complex human-technology relationships and interactions, deluge of data related to different aspects of human dynamics, and transdisciplinary challenges that involve natural sciences, social sciences, humanities, and engineering. It is not feasible for one edited volume to cover all of these issues. This edited volume instead is intended to contribute to human dynamics research through a collection of papers which focus on selected innovative approaches, data issues, method development, and empirical studies mainly from geographic and spatiotemporal perspectives. A Symposium on Human Dynamics Research consisting of twenty-five paper and panel sessions was organized at 2016 American Association of Geographers (AAG) annual meeting in San Francisco, California. An open call for papers was announced to a wider research community. The chapters in this edited volume are selected contributions from the Symposium and the open call.

1.2 Human Dynamics in Perspective

A legitimate question to ask at the beginning of this edited volume is: what is the precise meaning of human dynamics? Human dynamics have been studied in many different disciplines such as business, geography, physics, planning, psychology, sociology, among others. Each discipline tends to have its own perspective that leads to somewhat different research focuses. Jay Forrester, who was recognized as the founder of system dynamics, published three well-known books—*Industrial Dynamics* (Forrester 1961), *Urban Dynamics* (Forrester 1969), and *World Dynamics* (Forrester 1971). Forrester discussed computer simulation models in these books to shed light on the interactions among various elements for industrial management, urban issues, and the world system, respectively. Although the computer models proposed by Forrester offered useful tools for evaluating different strategies and policies, they were criticized for using a "systems analysis" approach to simulating social problems. For example, the computer simulation in the *Urban Dynamics* book was criticized by not considering that short-term gains might be desirable even at the price of long-term loss (Hester 1970). This suggests some shortcomings of using system dynamics approach to addressing social policy issues. The System Dynamics Society describes system dynamics as "a computer-aided approach to policy analysis and design. It applies to dynamic problems arising in complex social, managerial, economic, or ecological systems— literally any dynamic systems characterized by interdependence, mutual interaction, information feedback, and circular causality." (http://www.systemdynamics.org/what-is-s/) Human dynamics appear to be covered under this definition despite the weaknesses of tackling social issues using a system dynamics approach.

Wikipedia suggests that "Human Dynamics as a branch of statistical physics: Its main goal is to understand human behavior using methods originally developed in statistical physics." (https://en.wikipedia.org/wiki/Human_dynamics) This definition reflects a particular research direction of human dynamics that was inspired by Barabasi's (2005) publication of "The origin of bursts and heavy tails in human dynamics" in *Nature*. This article points out that most human dynamics models in various disciplines assume that human actions are randomly distributed in time. However, Barabasi's (2005) study indicates that individuals often execute tasks based on some perceived priority with bursts of rapidly executed tasks separated by long periods of inactivity, which results in heavy-tailed distributions. This line of research has influenced some human dynamics studies by geographers and geographic information scientists.

Seagal and Horne (2003), on the other hand, suggest that "Human Dynamics is a body of work that identifies and illuminates innate distinctions in the way people function as whole systems that include mental, emotional, and physical dimensions." The mental principle is related to mind and expressed in thinking (i.e., rational). The emotional principle is about forming relationships and expressed by making connections (i.e., relational). The physical principle is the practical part and expressed by doing and making things (i.e., pragmatic). All three principles are active in all people with various combinations and to varying degrees. This perspective of human dynamics places an emphasis on individual personality.

Human dynamics is a less frequently used term than other closely-related terms such as human interaction, human mobility, or human movement in geographic literature. One early use of "human dynamics" is found in Finch's article of "Geographic science and social philosophy" in the *Annals of the Association of American Geographers*, which states that "It is obvious that many of the elements of regions are observable things. This is important, because observation, as Krebs puts it, 'is and remains the basis of geography.' To be sure, the activities and forces of human dynamics, in which some would see the essentials of regional unity, are not all amenable to direct observation, but we gain awareness of them by only slightly different means. They are recognized by all regional geographers." Finch (1939, pp. 14–15) This article mentions "forces of human dynamics" that are not amenable to direct observation. It was very challenging in the 1930s to observe not only the forces of human dynamics but also human dynamics themselves due to the lack of tools enabled by technologies to collect such data, especially at a large scale. Traditional interview and survey methods are costly and time-consuming to collect and record human activities and interactions, which in turn present an impediment to applying theoretical frameworks to examining human dynamics at a community or society level. One example is Hägerstrand's time geography, which offers a useful framework for studying human dynamics under various types of constraints in a space-time context (Hägerstrand 1970, 1982). However, the basic concept of space-time path in time geography, which requires data collection of spatial movements over time (i.e., trajectory) for each individual, presents a major challenge to researchers in the era before the Global Positioning System (GPS) technology. Furthermore, even if researchers in the 1970s were able to collect trajectory data of 5000 individuals, it would be a daunting task to show all 5000 space-time paths with pen and paper. Time geography, which has become one of the most widely used conceptual framework in human dynamics research, is very much promoted by the advances of location-aware, mobile, and information and communication technologies during the last few decades that have removed major obstacles of data collection and analysis (Shaw 2012; Shaw et al. 2016). In the meantime, human dynamics research is gaining momentum in geography while geography as a discipline is increasingly recognized by researchers in other disciplines for providing an important perspective to human dynamics research.

Apparently a cogent definition of human dynamics that everybody can sign up for is still elusive due to the diverse disciplines and approaches in the rapidly evolving field of human dynamics research. Instead of defining the boundary of this evolving field, we believe that it is more productive to outline the core elements. The concepts, methods, and applications of human dynamics research are likely to evolve with the changing environments, technologies, and human societies. Research paradigms and dominant research methods employed today will be different down the road. For example, what geographers did 100 years ago are very different from what geographers do today in many aspects. There have been many debates of "what is geography?", yet the field of geography continues to evolve and move forward without a universal definition of the field. Nevertheless, we should not pursue human dynamics research without some directions and guidelines. Below are some of the core elements that should be considered in human dynamics research in our opinion.

- From "physical space" to "virtual space": With the modern technologies, human activities and interactions have been increasingly taking place in cyberspace, communication space and online social space (e.g., e-shopping, e-commerce, e-education, e-government, email, text messages, online social networks, online games). These activities and interactions in virtual space are not independent from human activities and interactions in physical space. In fact, they interact and influence each other in most cases. It therefore is critical to examine human dynamics in both physical and virtual spaces and their mutual interactions if we want to better understand how human dynamics are evolving and what smart and connected communities should be to better serve future human dynamics needs. Indeed, the on-line and off-line human activities are more closely coupled than ever.
- From "historical" to "real-time": Understanding changes of human dynamics over time (i.e., historical and long term) is as important as understanding human dynamics at this moment (i.e., real-time and short term). They are for answering different research questions and supporting different policy decisions. Observations of human dynamics at different temporal scales could lead to different findings about human dynamics. We need to examine human dynamics not only at various temporal scales but also the interactions among human dynamics at different temporal scales. The improved granularity of human activities will lead to a better understanding of human dynamics in space and time.
- From "human" to "context": Human activities take place within a context. Human dynamics research is not just about human. The environments and situations are important factors that influence human behaviors and dynamics. Recent research interests in place and semantics are good examples of deriving meanings behind human dynamics based on the context. There are at least three types of context that need to be considered in studying human dynamics. The first type of context is "what are around us?" which are usually recognized by an individual using relative locations in space (i.e., *relative space*). The second type of context is "what are related to us?" which are usually comprehended by people as relations among different entities (i.e., *relational space*). The third type of context is "what people have in mind" such as motivation, goal, perception, etc. (i.e., *mental space*). All three contexts can involve human and non-human elements. Non-human elements can be entities in the physical space (e.g., a road or a restaurant) or something in the virtual space such as the Google search website. On the other hand, traditional maps and geographic information systems (GIS) are based on absolute locations in space (i.e., *absolute space*). It is important to develop a theoretical framework that can integrate all elements relevant to human dynamics in absolute space, relative space, relational space,

and mental space in order to gain more comprehensive insights on the processes behind human dynamics beyond the observed spatiotemporal patterns.

Obviously, human dynamics is a slippery term that is hard to define its scope with one single definition. This section reviews several different perspectives of human dynamics research and suggests an approach of pursuing human dynamics research by focusing on some important concepts such as space, time, context, process, relationship, and interaction related to human dynamics. Development of a framework that integrates human dynamics in absolute space, relative space, relational space, and mental space can be very helpful to the human dynamics research community with different perspectives across various disciplines.

1.3   Overview of the Chapters in This Volume

This book is organized into 13 chapters. This chapter is written by the editors to provide the context and present an overview of all the chapters in this volume— human dynamics research in smart and connected communities. The subsequent substantive chapters cover various topics related to human dynamics research. This book ends with another chapter by the editors to discuss the limitation of this book as well as outlook and next steps of human dynamics research in the context of smart and connected communities.

Chapter 2 (Thakur et al. 2018) argues that the increasingly available geo-located data sources make it possible to understand human dynamics that previously was not possible. It presents four case studies of using geo-located cellphone data or social media data to improve land use classification, examine population dynamics of a major sport event, investigate transient population dynamics, and assess facility popularity to support its argument. The authors present results of using various methods on different types of data for classifying human population distribution, land use and facility popularity. Despite the data limitations in veracity and completeness, this chapter demonstrates a strong case of using geo-located data to gain insight into human dynamics at a fine resolution.

The research community has used many different types of tracking data to investigate various kinds of human dynamics. One of the common questions is potential biases embedded in each dataset for studying human dynamics. Chapter 3 (Xu et al. 2018) explores this issue based on call detail records (CDR) data and a more complete dataset that includes both CDR data and additional cellphone activities tracked by a cellphone service provider. One key finding of this study is that the number of active cellphone users is a better indicator of the spatiotemporal distribution of cellphone users than the volume of phone calls/text messages. This is consistent with a "burst" human activity pattern identified by Barabási (Barabási 2005); i.e., many human activities exhibit bursts of rapidly occurring activities separated by long periods of inactivity. Use of CDR data to study certain spatiotemporal human activity patterns therefore could be questionable since CDR data reflect where people initiate and/or receive phone calls and text messages rather than where people are distributed. In other words, CDR data could be biased from both spatial and temporal perspectives, especially for those people who use their mobile phones infrequently.

With the increasing human activities and interactions taking place in virtual space, it is critical to be able to represent, analyze and visualize human dynamics in both physical space and virtual space. Chapter 4 (Gao et al. 2018) proposes a spatiotemporal network framework to deliver such functions. It introduces physical edges for movements in physical space, social edges for social relationships and interactions, and physical-social edges to connect physical locations with their associated social activities. This study shows a case study for visualizing such a spatiotemporal network of geo-social interactions with Twitter data, followed by a discussion of four potential quantitative measures of complex interactions in the proposed spatiotemporal network. It is an example of integrating absolute space and relational space discussed in Sect. 1.2 above.

Trajectory data show the locations of moving objects over time that can be useful for studying spatiotemporal movement patterns. However, one major shortcoming of most trajectory data is the lack of semantic data associated with various locations. One popular research topic in recent years therefore is on deriving and managing semantic trajectory data. Chapter 5 (Fan and Stewart 2018) proposes a semantic data modeling framework that employs semantic web technologies to represent, query, reason, and visualize human movements. It builds an ontology-driven knowledgebase to integrate spatial, temporal and semantic data and also presents a use case of student movements on a university campus based on class schedules. This chapter demonstrates some innovative ways of working with semantic data beyond what traditional GIS data models can deliver.

Chapter 6 (Xu 2018) uses cybernetics as a framework to examine synergy between people and technology that transforms each individual and creates cyborgs. Due to increasing mix of activities and interactions in both physical and virtual spaces, this paper chooses Twitter data as an example to illustrate the complexity of cyborg identities. Cyborg, which is a hybrid of part organism and part machine, is used as an overarching identity concept in the Twitter world to help us address the challenge. Cybernetics in this paper helps formalize the relations between cities and their dwellers as communication and feedback loops. Cybernetics therefore can serve as a theoretical foundation to critically examine the technological means for achieving smart cities.

Human dynamics play an important role in many application areas, including many of our public health challenges. Chapter 7 (Wen et al. 2018) proposes a location-based client-server framework, which consists of a client-side smartphone-based risk assessment module and a server-side epidemic simulation model, for assessing personalized exposure to the risk of respiratory disease transmission. This paper represents an application of linking the dynamic movements of an individual to the potential of being exposed to the risks in surrounding environments. By keeping individuals informed about potential risk levels, such an application could influence individual behavioral patterns that reflect important interactions between information flows in virtual space and human movements in physical space.

As individual tracking data at high spatial and temporal resolution levels become increasingly available, privacy protection has been a challenging issue to deal with. With the modern technologies and big-data orientation, individual data are constantly collected by both private firms and government agencies. It is no longer realistic to assume that we have control over our data. It is a matter of who owns what data and how the data are being used. On one hand, the research community is hungry for detailed individual-level data to gain insights on human dynamics. On the other hand, research ethics mandate privacy protection. Chapter 8 (Miller and Hoover 2018) tackles this issue by measuring the uniqueness of locations associated with individual trajectories (i.e., unicity) based on a subset of GPS trajectories from the Microsoft GeoLife dataset. By exploring how unicity varies with the number of randomly selected points, temporal and directional information, and transportation modes, the findings suggest significant privacy concerns due to a high unique level of individual trajectories.

There are an increasing number of studies on the interplay between online social networks and geography to gain insight on the relationships between information flows in virtual space and locations in the real world. Chapter 9 (Koylu 2018) analyzes reciprocal conversations among individuals based on geo-tagged tweets in the U.S. to find out how the semantics of information vary based on the geographic locations and communication ties among the users. This study proposes an approach of using spatial network smoothing and probabilistic topic modeling to extract geo-social semantics that reflect geo-social dynamics of the society.

As the research community is gaining momentum on paying attention to place besides space, Chapter 10 (Cheng and Shen 2018) extends the authors' previous work on "where, when and how long you stay is who you are" to "what place, when and how long you stay is who you are" by shifting from a focus on space-time activity patterns to a focus on place-time activity patterns. This study uses London's police foot patrol tracking data to demonstrate the proposed approach and methods. The results indicate that police who patrolled different locations in London could share a similar place-time activity pattern because different locations are associated with the same semantic meaning. This introduces a new way of measuring similar space-time behavioral patterns.

With an increasing interest in the interactions between virtual space and physical space and a lack of tools for exploring such interactions, Chapter 11 (Ye et al. 2018) presents a Social Network Simulator with functions supporting network generator, network analysis, community detection, and information diffusion modules in an open source package for exploring information diffusion patterns in a social network over time, especially for spatial meme diffusion. This chapter suggests a need of further developing open-source tools to support researchers who study human dynamics that require data management, analysis, and visualization functions beyond what traditional toolkits can offer.

Chapter 12 (Nara et al. 2018) discusses some challenges and opportunities of using social media and big data for human dynamics research. This chapter uses the papers presented in nine paper sessions organized at 2016 and 2017 annual meetings of the American Association of Geographers (AAG) to summarize the data, methods, and applications reported in those papers. The results indicate that a wide range of data, methods and applications have been investigated under the broad human dynamics theme. In addition to seven research challenges that were reported in the literature before, this chapter suggests that frequent changes of the ways that online social media data can be accessed by researchers and data/ algorithm uncertainty as two new challenges to human dynamics research.

In the final and concluding chapter (Sui and Shaw 2018), the editors recap the major findings, identify the gaps of the literature, and outline future research directions related to human dynamics in the broader context of smart & connected communities. In particular, the editors emphasize the importance of integrating organic and designed data, crossing the chasm of quantitative and qualitative approaches, and balancing the positive and normative dimensions. Future research on human dynamics in the context of smart & connected community should focus not only on efficiency, but also on equity and sustainability. Last but not the least, the editors challenge the human dynamics research community to embrace the open science paradigm to make all our future research reproducible, replicable, and generalizable. This is the only way to maintain the momentum to make human dynamics research more robust and reliable.

The collection of papers in this volume covers selected topics in human dynamics research, especially from data-driven and analytical perspectives. This orientation reflects the background of the contributing authors whose research interests mainly focus on the analytic aspects of geographic information science and geography. As the first volume in the *Human Dynamics in Smart Cities* book series by Springer, this edited volume serves as a useful reference for the data-driven and analytics side of human dynamics research community. We anticipate additional volumes down the road to cover other perspectives of Human Dynamics in Smart Cities.

References

Barabási, A.-L. (2005). The origin of bursts and heavy tails in human dynamics. *Nature, 435* (7039), 207–211.

Cheng, T., & Shen, J. (2018). *Grouping people in cities: From space-time to place-time based profiling.* (This volume).

Fan, J., & Stewart, K. (2018). *Modeling mobility and dynamics of scheduled space-time activities —An RDF approach.* (This volume).

Finch, V. C. (1939). Geographical science and social philosophy. *Annals of the Association of American Geographers, 29*(1), 1–28.

Forrester, J. (1961). *Industrial dynamics.* Cambridge, MA: MIT Press.

Forrester, J. (1969). *Urban dynamics.* Cambridge, MA: MIT Press.

Forrester, J. (1971). *World dynamics*. Cambridge, MA: Wright-Allen Press.

Gao, S., Chen, H., Luo, W., Hu, Y., & Ye, X. (2018). *Spatiotemporal-network visualization for exploring human movements and interactions in physical and virtual spaces*. (This volume).

Hägerstrand, T. (1970). What about people in regional science? *Papers of the Regional Science Association, 24*(1), 7–21.

Hägerstrand, T. (1982). Diorama, path and project. *Tijdschrift voor Economische en Sociale Geographie, 73*(6), 323–339.

Hester, J., Jr. (1970). System analysis for social policies (Book Review: Urban dynamics by Jay W. Forrester. M.I.T. Press, Cambridge, Mass., 1969. xiv + 290 pp.). *Science, 168*(3932), 693–694.

Koylu, C. (2018). *Uncovering geo-social semantics from the Twitter mention network: An integrated approach using spatial network smoothing and topic modeling*. (This volume).

Manyika, J., Chui, M., Brown, B., Bughin, J., Dobbs, R., Roxburgh, C., et al. (2011). *Big data: The next frontier for innovation, competition, and productivity*. USA: McKinsey Research Institute.

Miller, J., & Hoover, B. (2018). *An exploratory analysis of the effects of spatial and temporal scale and transportation mode on anonymity in human mobility trajectories*. (This volume).

Nara, A., Tsou, M.-H., Yang, J.-A., & Huang, C.-C. (2018). *The challenges and opportunities with social media and big data for research in human dynamics*. (This volume).

Seagal, S., & Horne, D. (2003). Human dynamics for the 21st century. *The Systems Thinker, 14*(1), 2–6.

Shaw, S.-L. (2012). Guest editorial introduction: Time geography—Its past, present and future. *Journal of Transport Geography, 23*, 1–4.

Shaw, S.-L., Tsou, M., & Ye, X. (2016). Editorial: human dynamics in the mobile and big data era. *International Journal of Geographical Information Science, 30*(9), 1687–1693.

Sui, D., Elwood, S., & Goodchild, M. (Eds.). (2013). *Crowdsouring geographic knowledge: Volunteered geographic information (VGI) in theory and practice*. Dordrecht, The Netherlands: Springer.

Sui, D., & Shaw S.-L. (2018) *Outlook and next steps: From human dynamics to smart and connected communities*. (This volume).

Thakur, G. S., Sims, K. M., Mao, H., Piburn, J. O., Sparks, K. A., & Weber, E. M., et al. (2018). *Utilizing geo-located sensors and social media insight for research in population dynamics and land classification*. (This volume).

Wen, T.-H., Hsu, C.-S., Sun, C.-H., Jiang, J.-A., & Juang, J.-Y. (2018). *A location-based client-server framework for assessing personal exposure to the transmission risks of contagious diseases*. (This volume).

Xu, C. (2018). *Smart sensors, cyborgs, and cybernetics: A critical reading of smart city technologies*. (This volume).

Xu, Y., Shaw, S.-L., Lu, F., Chen, J., & Li, Q. (2018). *Uncovering the relationships between phone communication activities and spatiotemporal distribution of mobile phone users*. (This volume).

Ye, X., Dang, L., Lee, J., & Tsou, M.-H. (2018). *Open source social network simulator focusing on spatial meme diffusion*. (This volume).

Author Biographies

Shih-Lung Shaw is Alvin and Sally Beaman Professor and Arts and Sciences Excellence Professor of Geography at the University of Tennessee, Knoxville. He also serves as the Interim Associate Provost for international education at the University of Tennessee, Knoxville. He received his B.S. degree from the National Taiwan University and his M.A. and Ph.D. degrees from the Ohio State University. His research interests cover geographic information science (GIScience), transportation geography, time geography, GIS for transportation (GIS-T), and space-time analytics of human dynamics. His recent research has focused on space-time analytics of human activities and interactions in a hybrid physical-virtual world based on various types of individual tracking data such as cell phone data, online social media data, vehicle tracking data, travel-activity survey data, and population migration data. His research has led to the development of a space-time GIS for representation, analysis, and visualization of individual activities and interactions in a hybrid physical-virtual space. Dr. Shaw is a Fellow of the American Association for Advancement of Science (AAAS). He also received the Edward L. Ullman Award for Outstanding Contributions to Transportation Geography from the Association of American Geographers (AAG) and served as the Head of the Department of Geography at the University of Tennessee, Knoxville.

Daniel Sui is Arts and Sciences Distinguished Professor and Professor of Geography, Public Affairs, Public Health, and Urban/Regional Planning at the Ohio State University (OSU). Since July 2016, he has been on an IPA assignment to serve as the Division Director for Social and Economic Sciences (SES) at the U.S. National Science Foundation. Prior his current appointment, Daniel Sui served as Chair of Geography (2011–2015) and as Director of the Center for Urban & Regional Analysis (CURA) (2009–2012) at OSU. He holds a B.S. (1986) and M.S. (1989) from Peking University and Ph.D. from University of Georgia (1993). His current research interests cover issues related to robust and reliable practices in spatially integrated social sciences and humanities, human dynamics in smart & connected cities, and location-based social media. Daniel Sui was a 2009 Guggenheim Fellow, 2014 Google Faculty Fellow, and 2014 recipient of the distinguished scholar award from the Association of American Geographers. Sui was also the 2015 Public Policy Scholar in residence at the Woodrow Wilson International Center for Scholars. More information about Sui's current research can be found at: https://www.researchgate.net/profile/Daniel_Sui.

Chapter 2
Utilizing Geo-located Sensors and Social Media for Studying Population Dynamics and Land Classification

Gautam Thakur, Kelly Sims, Huina Mao, Jesse Piburn, Kevin Sparks, Marie Urban, Robert Stewart, Eric Weber and Budhendra Bhaduri

2.1 Introduction

Since the Sumerians started using clay beads to record trade some 7000 years ago to the invention of the tabulating machine in the late 19th Century, our ability to generate data has always, and will continue to, outpace our ability to record and analyze it. This perpetual race sees a continual pattern of breakthroughs in the ability to collect data followed by a period of time when new methods are developed to help make sense of this new source of information. Over the past decade data collection has pulled ahead once again, particularly data on human activity and locations. Thanks in large part from the maturation of mobile sensors and the proliferation of user-generated content like social media, it's estimated that 90% of the data generated since the dawn of civilization has been created in the last two years alone.

With increasing sources of data on human dynamics comes a new ability to understand human dynamics that previously was not possible. Geo-located user generated content from mobile devices allow a level of spatial and temporal granularity that would be prohibitively expensive and time consuming for more traditional methods such as surveys and censuses. The ability to understand where, when, and why humans move across space and time has always been essential to research areas such as urban planning, transportation, population dynamics, and emergency preparedness and response. Geo-located user generated content is allowing new insights into these and many other fields.

Geo-located user generated content can come in several forms. The spatiotemporal GPS locations of cell phones is one form that has demonstrated the ability to shine light on how populations move across and shape the built environment and

G. Thakur (✉) · K. Sims · H. Mao · J. Piburn · K. Sparks · M. Urban · R. Stewart
E. Weber · B. Bhaduri
Oak Ridge National Laboratory, Oak Ridge, TN, USA
e-mail: thakurg@ornl.gov

© Springer International Publishing AG, part of Springer Nature 2018   13
S.-L. Shaw and D. Sui (eds.), *Human Dynamics Research in Smart and Connected Communities*, Human Dynamics in Smart Cities, https://doi.org/10.1007/978-3-319-73247-3_2

more recently the use of social media on mobile devices has shown tremendous potential to allow us to better understand the distribution of populations at a specific point in space-time. Beyond that, social media data has enabled researchers to understand population activities, events, and underlying cause that generates the dynamics of population movements.

Still challenges remain in our ability to effectively put this data to use. Before one can ask questions of the data we must first collect it and the sheer volume and variety of geo-located user generated content presents a challenge in the gathering, storing, and querying of the data. This majorly occurs because of velocity with data gets generated, validating the data, and the sheer volume of the data. Only once the data is collected, cleaned, and formatted can we begin the next challenge of developing insightful methods that allow us to ask questions we previously were not able to answer.

Four case-studies of using geo-located user generated for human dynamics research are provided. These studies are inspired from the work of Stewart et al. (2017) while initiating a research agenda in human dynamics and land use domain by proposing an explicit model that assists in delineating and articulating the opportunities, challenges, and limitations of using social media. In the first study, mobile phone call volume and GPS locations are used to characterize human activity patterns and provide inference on land use in Dakar, Senegal. Next, we demonstrate the ability of geo-located social media posts to provide insight on population density estimates for special events, such as sporting events or emergency situations. An example of this is presented with college football games on a university campus. Finally, we provide an initial investigation on the potential to identify a specific building's facility use type based solely on geo-located social media content. These studies demonstrate innovative findings that are only made possible because of new forms of data about human activity and their mobility patterns. The chapter underpins the need to utilize new forms of data collection mechanism as well as their use to augment our understanding of human dynamics research and future application of geographical information systems.

2.2   Geo-located Human Activity Data Collection and Management

User generated content with locational metadata is a captivating attribute for modeling population, fine-resolution land use and land cover classification. The spatio-temporal feature, included with potentially descriptive user activity, can provide improved ways of modeling population dynamics at high-resolutions. Additionally, researchers can begin to understand better ways to approach challenges involved in the medical, political, historical, environmental, social, and technological fields (Kuhn 2012). The instantaneous spatial information, much of which is provided by everyday citizens, provides a deeper understanding in how land use topology changes and maintain at different scale and sizes of settlements. This section provides background in the development and current status of geographically focused user-driven content harvested through crowd-sourcing, sensors, and social media.

2.2.1   Neogeography

The evolution from a one-way web browsing experience (e.g. Web 1.0) to now, a two-way interaction and sharing medium is referred to as Web 2.0. Some believe this transformation was so revolutionary in its earliest stages that the term 'Neogeography' was born to encapsulate a new era of geographic practices through web-based operations (Hudson-Smith et al. 2009). One of the most influential instruments to Neogeography was the public release of a free, web-based mapping interface, known as Google Earth. In lieu of expensive cartographic and GIS applications, anyone with access to the Internet could easily upload and share geographic data to make their own maps. This in turn ignited an interest of dynamic mapping to the everyday masses (Turner 2006). Since then, computer scientists have begun constructing other mapping applications and interfaces to produce maps that are less labor intensive and more automated.

2.2.2   Volunteered Geographic Information

Innovations in mobile Internet access (i.e. smartphones) and authority (i.e. non-experts) are redefining how the sciences are adopting non-traditional data. This is a result of the Internet morphing into a modernized power structure of bottom up practices, and what Michael Goodchild termed volunteered geographic information, or VGI (Goodchild 2007). For example, it has become common practice to rely on smartphones to browse the Internet for quick information about a place. And, it is also not uncommon for that data we are searching for to be provided by our peers, rather than established experts. Instead of the traditional construction of knowledge, produced and edited strictly by trained professionals, most anyone with an Internet connection now has the ability to share their local knowledge, and in the case of VGI, with locational attributes. Geographic information disseminated by citizens ultimately becomes a new avenue of information delivery and consumption. Additionally, diverse human contributors can facilitate unique and/or specific geographic knowledge with the simplest post of a tweet (Twitter), status update (Facebook), or picture (both).

2.2.3   Social Media

The enormous popularity of social media suggests that user generated content is here to stay, and potentially, is an acceptable lifestyle of continuous information exchange. With boundless content, users share excerpts of their whereabouts, opinions, beliefs, activities, etc., for the sole purpose of broadcasting one's life to interested parties (Grace et al. 2010; Java et al. 2007). While they may not be intentionally sharing their experiences in hopes of providing data for science, they are nevertheless publicly sharing their activities to socially connect (Croitoru et al. 2012, 2013). Goodchild (2007) has referred to this population as "Citizen Sensors." Their chaotic and unsystematic availability of information proves worthwhile for application program interfaces (APIs) which can pragmatically mine user generated content (e.g. Tweets, Check-ins) to give a quantitative look into the world around us. This chaotic and unsystematic availability of user information is what has provoked a relationship between scientists and Big Data.

2.2.3.1   Twitter

As of October 2016, at least 550 million tweets were sent per day from the 100 million daily active users (Twitter IPO filing). With this plethora of streaming information from certainly one of the industry's leaders, Twitter is an ideal platform source to harvest media feeds from through its API. On the company's website, their policy states that in exchange for using their services, the user agrees to have their information made public and searchable by third parties, as long as the user's account privacy settings are set accordingly (Twitter Privacy Policy). This agreement is what allows this research to explore Twitter's 140-character messages to better understand actual, first-hand experiences during a live event. Additionally, the fact that more than three-quarters of active users socially connect through a mobile device only further supports the possibilities of gaining up-to-date access with location accuracy (Twitter IPO filing). However, social media users are a sample of the general public since not everyone participates. Nevertheless, any amount of publicly shared experiences or actions, recorded in the moment through smartphones, especially those with locational tracking services activated, can ideally illustrate a more accurate depiction of population dynamics.

2.2.3.2   Facebook

In its origin in 2004, Facebook was only accessible to college students from a few selected universities. Today, Facebook connects individuals (13 years of age or older), all across the world with 1.13 billion daily active users.[1] While Facebook

---

[1]http://newsroom.fb.com/company-info/. Retrieved: September 14, 2016.

may not be the microblogging platform that Twitter is, user can tag to their post unique venues thereby creating a map-blog of where users have been and when. This geolocation attribute available from Facebook has made check-ins an attractive source of data for researchers. Unfortunately, such detailed recordings of a user's activities and location often times prompt stricter privacy concerns, which leads to more challenging harvesting techniques. The first step Facebook has taken to limit these infractions is for Facebook's Public Feed API to be unlike Twitter's in that only a few selected media agencies can stream status updates (Public Feed API). There is no free API service available to the public. These digital barricades are difficult to overstep, which is why the next best spatial and temporal clues to attendee activities are to monitor Facebook Place check-ins. By tracking check-ins to location-specific Facebook Places, a spatial flow of these populations could be interpreted defining movement patterns or local hotspots. Unfortunately, check-outs are not usually provided from users, therefore these patterns only suggest when and where areas experience increased activity.

2.2.3.3 Limitations of Social Media

Data may not always be coherent or spatially accurate due to the very nature of user generate content. As non-professionals, these users are providing large amounts of data that are "inherently noisy" (Becker et al. 2010) which requires supplementary efforts to interpret. According to Elwood et al. (2012), because a large amount of unstructured data is generated, it also "presents a number of challenges for developing methodologies." Without an authoritative structure to restrict erroneous observations, information can produce false-positive results skewing representations of population dynamics. Tweets are notorious for this in their very nature due to such limited posts that can be incorrectly interpreted or, more importantly, untruthful. It was assumed in this research that the intentions of users were to honestly share information about their current activities, at the appropriate moment in time.

2.2.3.4 Application to Human Dynamics Research

Several of these datasets have proved to be very useful in understanding the dynamics of human mobility. In this chapter, we will focus the use of cell phone call pattern, Twitter activity as well as Facebook graph data to augment our understanding of population dynamics and demonstrate their efficacy and use in the changing world of geographic information sciences.

2.3   Improving Land Use Inference by Factorizing Mobile Phone Call Activity Matrix

Land use is "the human employment of the land" (Meyer and Turner 1996) and characterized by "the arrangements, activities and inputs people undertake" (Di Gregorio and Jansen 1998). Depending on the socioeconomic functions, land uses can be categorized into industrial, commercial, residential, administrative, agricultural, etc. Understanding land use is crucial for practitioners and researchers to perform urban analysis and planning, such as population estimation, infrastructure planning, neighborhood zoning, and hazard and pollution analysis (Donnay and Unwin 2001). Traditionally, land use is identified through surveys, which usually tend to be expensive and difficult to obtain frequently. In resource-constrained environments, where surveys are usually rare, detailed land use data is scarce, if not missing. In order to attack this problem, we propose to detect land use based on human activity patterns. The idea of our work lies on the fact that land use is shaped by human activities. Novel data sources that record the detailed footprint of human activities are enabling new methods for land use inference. Social media data and Internet access activity data, for instance, have been used to measure land use in cities (Hudson-Smith et al. 2009) and building use (e.g. residential buildings, research labs, auditorium) on campus (Kuhn 2012). Although the Internet penetration rate is low in poorest countries, the adoption of mobile phones is becoming nearly ubiquitous: there are 4.7 billion unique mobile phone users worldwide (http://www.gsma.com/mobileeconomy/). Several studies investigated the potential of mobile phone data for spatio-temporal population distribution modeling, and applied clustering, classification, and Eigen-decomposition methods to identify call patterns for land use detection (Frias-Martinez et al. 2012; Reades et al. 2009). Yet, most existing studies focus on developed countries, probably due to the extensively deployed sensors and rich ground-truth data available. In resource-constrained countries, there is far less geotagged human activity data, and detailed land use data is also scarce, if not missing. Moreover, different types of land uses are usually mixed and co-located in low-income countries, unlike land use zoning in developed countries. So, due to the limited data sources and mixed land use, it is unclear whether similar findings can be generalized to low-income countries. In this work, we examine the extent to which anonymized Call Details Record (CDR) data can be used to infer land use in Africa.

Our CDR data is provided by the Orange Telecommunications Company for the Data for Development Challenge-Senegal. This dataset contains the number and duration of calls between each pair of antennae (1666 antennae in total) on an hourly basis from 01/01/2013 to 12/31/2013. Since our study is based on an aggregated spatial analysis, users' anonymity is maintained. According to Telecommunication/ICT Development Report (2015), the mobile cellular subscriptions per 100 people in Senegal reached 99 in 2014. Given the high penetration rate, we consider mobile phone usage data as a good proxy for human activities. We process the CDR data in three steps in order to capture the spatiotemporal patterns of human activities. First, for each antenna, we aggregate its total call volume (including the number of incoming and outgoing calls) on an hourly basis, and normalize the original hourly call volume time series into z-score over the study period. Second, in order to capture the relative activity of a certain area at a given hour, we subtract the average activity (or baseline activity) of all areas during the hour from its normalized call volume. Third, we calculate the average of relative activity for the hour of day and day of the week, and obtain a 168-dimension (i.e. 7 days' times 24 h) call pattern vector for each cell tower. Combining the call pattern vectors of all cell towers, we have a call pattern matrix $V_{m \times n}$, where m is the number of cell towers (m = 477) and n is the length of a call pattern vector (n = 168). To decompose V, we apply the non-negative matrix factorization (NMF) techniques, which results in two matrices: weight matrix (**W**) and basis vectors (**H**). The mathematic formulation is $\boldsymbol{V_{m \times n}} \approx \boldsymbol{W_{m \times k} H_{k \times n}}$ where k is the number of basis vectors representing different land use patterns; and here we let k = 2. For the details of NMF, interested readers may refer to (Seung and Lee 1999; Berry et al. 2007). The implementation is based on the *scikit-learn* Python library.

2.3.1   Result and Analysis

Figure 2.1 shows two resulting basis vectors, denoted by Component 1 and Component 2. It can be seen that these two components show opposite patterns: peaks of Component 1 are in the day time (at 1 PM) and weekdays, while peaks of Component 2 are in the night (at 10 PM) and weekends, capturing commercial/business/industrial (C/B/I) and residential characteristics, respectively. Then, the original call pattern time series of each cell tower can be approximated by a linear combination of these two basis vectors based on weights contained in each row of W. Here, we classify an area as C/B/I if $w_1 > w_2$, and residential, otherwise. Voronoi polygons are used to approximate the reception area of cell towers (González et al. 2008). Based on the estimated call patterns for each tower, we infer the land use map, which is shown in Fig. 2.2a. The color value in the map is determined by the weights for two basis vectors, i.e. $w_1/w_1 + w_2$. Blue hue represents C/B/I, while red hue indicates 'residential' (or out-of-work) land use.

Voronoi polygons, however, are not actual neighborhood boundaries. In order to obtain a land use map in a physically meaningful context, we convert it to neighborhood level, based on the proportion of area intersections between cell tower's Voronoi polygons and neighborhood regions (see Fig. 2.2b and c).

2.3.2   Validation

In order to validate our results, we compare with ground-truth data. However, the official land use data in Dakar is missing. So, we introduce the OpenStreetMap

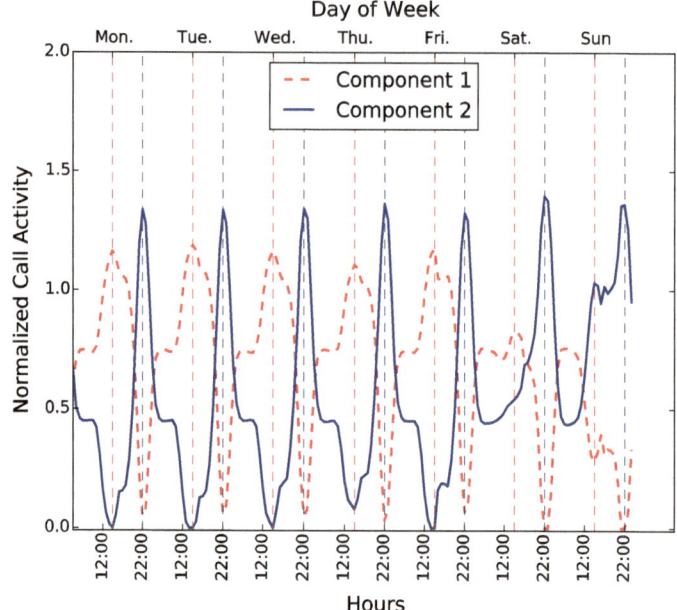

Fig. 2.1 Two basis vectors obtained from decomposing the 24-h/7-day call pattern matrix using non-negative matrix factorization

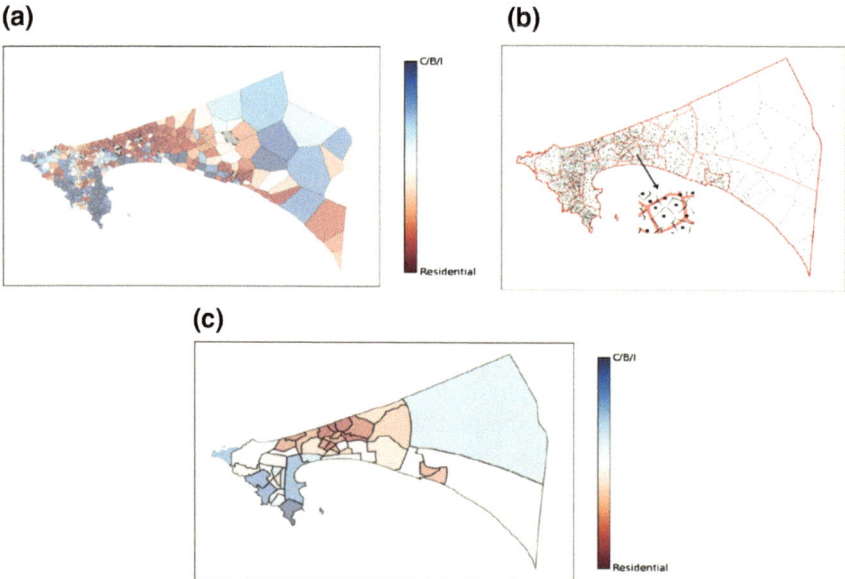

Fig. 2.2 Estimated land use map based on call activities: **a** at the voronoi polygon level, **b** intersection between voronoi polygon (grey dashed line) and neighborhood (red solid line) boundaries, and **c** at the neighborhood level

Table 2.1 Comparing POI landuse estimation ('commercial/office/industrial/residental/') from OSM

Predict	POI			
	'Commercial' (%)	'Office' (%)	'Industrial' (%)	'Residental' (%)
C/B/I	**90.4**	**76.6**	**63.6**	39.3
Residental	9.6	23.4	36.4	**60.7**

Points-of-Interest (POI) data as the ground-truth. POI data is voluntarily added by users to tag the function of a place, e.g. office, shop, leisure. Table 2.1 shows the comparison of two-class predictions with POI features, 'commercial/office/industrial/residential'. We find that a majority of 'commercial/office/industrial' areas are under the predicted C/B/I class, while over 60% of 'residential' places are within the predicted residential class. In addition, Fig. 2.3 shows the distributions of more POI features under these two classes. When $w_1 > w_2$, dominating features include 'embassy, garage, hotel, university, public building, restaurant', and 'military', whereas POIs, such as 'religious, cemetery, clinic, kiosk, farmland', and 'greenfield', take the larger proportions when $w_1 \leq w_2$. These findings are well aligned with the actual land use characteristics: for instance, embassies, public buildings, and offices are usually located in the commercial and business areas;

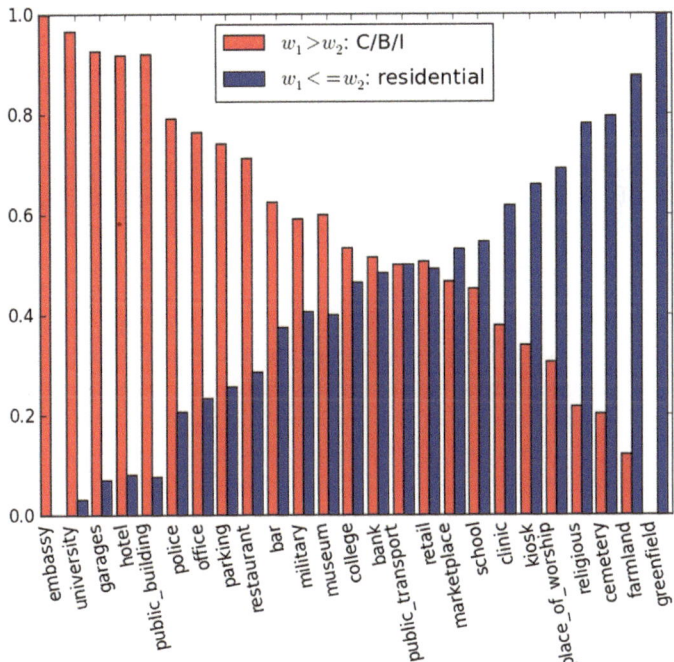

Fig. 2.3 Comparing land use estimation with multiple types of poi from OSM

religious places, clinics, and kiosks (i.e. very small shops) are usually located at residential neighborhoods. Also, cemeteries, farmland, and greenfield, which are often close to residential instead of work areas, are found to be more present when $w_1 \leq w_2$.

Overall, our results demonstrate the effectiveness of proposed method for land use detection, especially, when there is a 90.4% detection rate for 'commercial' area and 76.6% for 'office' area, but the detection rates for 'residential' (60.7%) and 'industrial' (63.6%) areas are relatively low. This is perhaps due to the fact that residential and industrial areas often co-locate with other land use types in Dakar. We acknowledge the limitation of this study is the lack of multiple types of land use classification. Future work needs finer-scaled human activity data and contextual information to further classify C/B/I into separate classes and improve mixed land use inference.

2.4 Understanding Special Events Population

Modeling population distributions at a high spatial and temporal resolution requires accounting for the dynamic nature of human populations. Models and representations of population that rely on census counts necessarily miss these dynamics. However, some recent efforts incorporate daytime or diurnal distributions (Kobayashi et al. 2011; Bhaduri et al. 2007), and episodic or tourist populations (Jochem et al. 2013; Charles-Edwards and Bell 2013) in order to better capture population dynamics. All of the efforts to extend population modeling beyond static representations of nighttime populations contribute to the overall goal of achieving what Martin et al. (2015) call "a full representation of population time," which is important for emergency preparedness and response, risk modeling, and many other applications. To continue these efforts, publicly available data feeds from social media offer an additional opportunity to improve population models at high resolutions (Bukhari et al. 2012; Birkin 2013).

As of late 2015, the world's two leading social media platforms are Twitter, with 320 million monthly active users, and Facebook, with 1.5 billion monthly active users. Only a portion of social media data has associated location information, however. For example, although an estimated 550 million tweets were sent per day as of October 2015, estimates of the portion of tweets that are geo-located have ranged from 0.47% (Cheng et al. 2010; Morstatter et al. 2013).

A natural application of the geo-located subset of social media data is the modeling of episodic populations associated with special events having high attendance and a significant presence on social media; in particular, this study focuses on game-day college football fans at The University of Tennessee (UT), Knoxville. Using tweets (from Twitter) and check-ins (from Facebook), this research integrates this new form of data in a high-resolution dasymetric population distribution model.

2.4.1   Methods and Results

The area within a 1.5-mile radius around The University of Tennessee football stadium was chosen for this study, and the population associated with football game-days was modeled. Geo-located tweets and check-ins were collected for the 24-hour period surrounding the scheduled kickoff for each home game in 2013. Seven terms associated with the university were used to filter tweets from Twitter's streaming API (Table 2.1). A cumulative count of Facebook check-ins was captured every 30 min for 95 establishments associated with game-day activities (e.g., restaurants and tailgating locations) (Table 2.2).

Two scenarios were modeled: (1) a "non-game-hours" scenario, and (2) a "game-hours" scenario. Each model outputs a population estimate for each cell in a raster grid with 3-arc-second resolution (~90 m). The 2012 version of the LandScan USA (Bhaduri et al. 2007) gridded nighttime dataset was used as a baseline population distribution to which the new modeled distributions could be added to create the final output grids. The LandScan USA nighttime distribution better represents the study area on a Saturday (all game-days were Saturdays) than the daytime dataset, because the daytime dataset assumes a weekday distribution of workers and students, which is very different from what would be expected in the area on a weekend.

A consistent measure of social media activity was required for each raster cell. First, the tweets and check-ins (collectively referred to as "posts" hereafter) were divided into two sets based on their timestamps. Posts were considered for the game-hours scenario if they occurred less than two hours prior to kickoff, or less than three hours after kickoff. All posts outside of those hours were considered for the non-game-hours scenario. A count of tweets and a count of check-ins were computed for each raster cell for each of the two scenarios, resulting in four raw count rasters.

Because of the limited amount of geo-located posts and the spatial errors in the associated location information, some locations that attract event populations may have no representation in the social media data. To overcome this limitation, kernel density estimation with a radius of two grid cells was performed on each of the

Table 2.2 The total seasonal count of game-day geocoded tweets associated with The University of Tennessee, 2013

UT term/phrase	Number of geocoded tweets
"Tennessee"	11,582
"Vols" (abbr. of "Volunteers", the team nickname)	7582
"GBO" (abbr. of "Go Big Orange", a common chant)	2495
"VFL" (abbr. of "Vol for life", a common slogan)	1612
"Neyland" (name of football stadium)	1144
"Football Time In Tennessee" (common slogan)	1135
"Big Orange" (a common alias for the team)	418

four-raw count rasters to estimate tweet densities and check-in densities across the grid. Then the densities were scaled so that each value represents posts per cell, and can be interpreted as an interpolated count.

For each scenario, a linear relationship between social media activity and event population is assumed. For each raster cell i, the special event population (y) is described by:

$$y_i = \beta_T w_i, \tag{2.1}$$

where β_T is a linear coefficient specific to the scenario T, and w_i is the number of posts at cell i. The β coefficient is the number of fans represented by each geo-located post. If available, an observed value or estimate of the total population in the study area associated with a scenario can be used to estimate β_T:

$$\beta_T = \frac{E_T}{\sum_{i=1}^{n} w_i}, \tag{2.2}$$

where E_T is the estimated total special event population for scenario T. The study area total for the game-hours and non-game-hours scenarios can be estimated by summing two separate components of the special event population: A, the ticketed fans, which is estimated by averaging the recorded attendance for each game, and α, all the other (non-ticketed) people in the area specifically for the event:

$$E_T = \lambda_T(A + \alpha) \tag{2.3}$$

where λ is a parameter representing the estimated portion of the peak game-day population. Ultimately, the final population estimate for each cell i is the sum of the baseline LandScan USA population (L_i) and the special event population (y_i):

$$P_i = L_i + y_i \tag{2.4}$$

In many special event situations, data often will not be available to support precise estimates of the parameters, α and λ, in Eq. 2.3. But event officials often have expert knowledge and are privy to information that allows reasonable estimates of these parameters. Ultimately, a software solution aimed at event officials allows such knowledge to be incorporated in the parameterization.

Figures 2.4 and 2.5 show example representations of the non-game-hours and game-hours scenarios, respectively, using rough estimates of these parameters. The estimate for α is 30,000, which is meant to include non-ticketed fans, city security, local business workers, and stadium staff, security, and teams. For the game-hours scenario, λ was set at 1, assuming the population peaks during the game. The non-game-hours scenario is meant to represent a moment approximately three hours before kickoff, for which λ equals two-thirds.

Large populations can be seen in and near the stadium in both Figs. 2.4 and 2.5, but with much greater concentration in Fig. 2.5. Figure 2.4 shows greater

Fig. 2.4 An example game-day population distribution around The University of Tennessee, Knoxville, during non-game hours: **a** Modeled event population, **b** Combined population (baseline + event population)

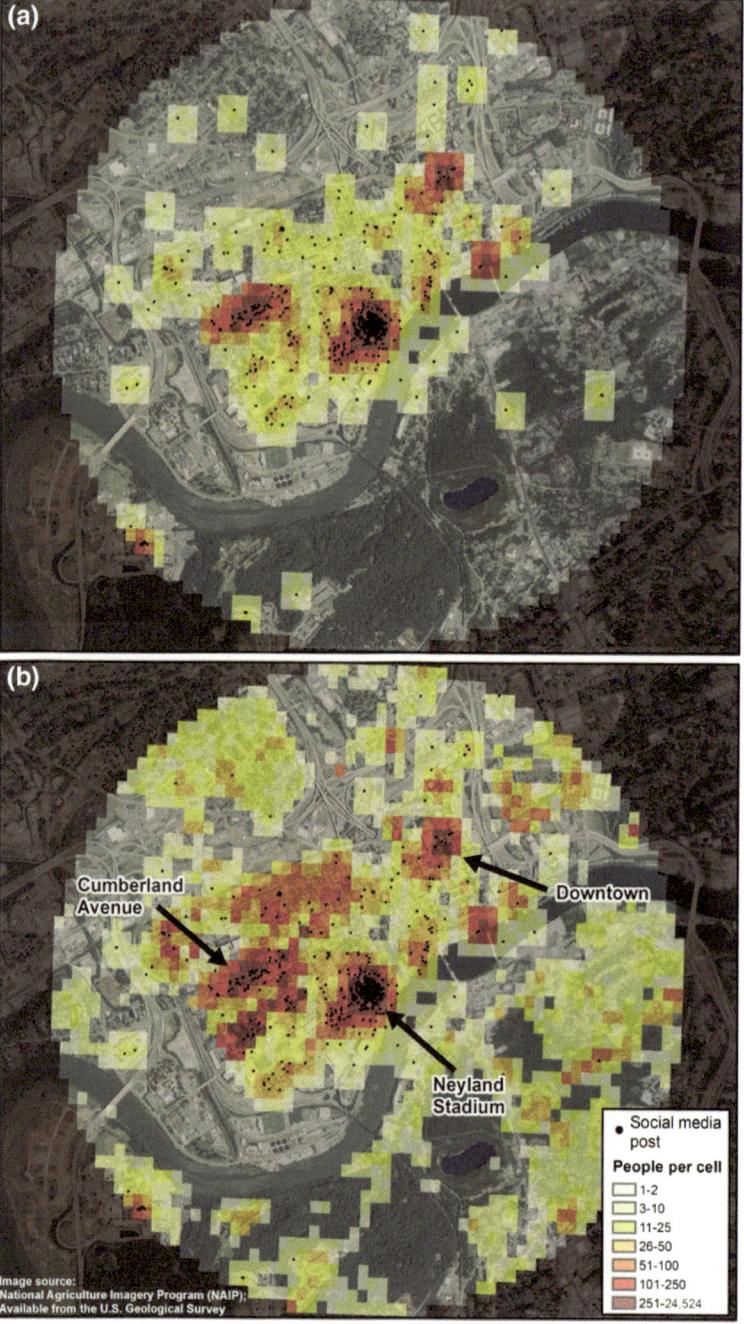

Fig. 2.5 An example game-day population distribution around The University of Tennessee, Knoxville, during game hours: **a** Modeled event population, **b** Combined population (baseline + event population)

concentrations in areas on and near campus that are popular for pre-game tailgate parties, as well as along Cumberland Avenue and in the downtown area, where restaurants, bars, and shops are concentrated. Different parameterizations (for α and λ) lead to different absolute population values; but the overall pattern and the ratios among the values remain the same (because the spatial distribution is based on only the social media data).

2.4.2 Discussion

The population distributions in Figs. 2.4 and 2.5 generally correspond with the spatial patterns familiar to the authors, and therefore suggest social media having a positive relationship with this particular special event population. However, greater quantities and greater spatial precision of observations of the population would be needed to test the assumed linear fit (or to fit alternative models). Greater quantities of data also would allow more refined temporal resolution, rather than having to aggregate data from across several football games into two general scenarios, as was done here.

How to estimate the total population present for an event also deserves future research. In the college football scenario, attendance data are helpful for estimating a portion of the total population, but the α parameter (non-ticketed event population) from Eq. 2.3 is more difficult to estimate. A scenario including ancillary data about the counts of subpopulations such as security personnel and event staff would have greater certainty. The data available to an analyst, and the analyst's familiarity with an event, play an important role in establishing reliable estimates of this parameter.

The filtering of the social media data is also a crucial step that relies on knowledge about an event and its location. The authors were able to eliminate a large subset of irrelevant social media data through search-term filtering of Twitter data and identification of relevant event-related establishments from the Facebook data. Of course, other events could prove more challenging because relevant search terms or relevant establishments could be unknown. Again, the development of a software solution that allows analysts with detailed knowledge about an event and location of interest to easily implement a model like the one demonstrated here would be a reasonable next step in expanding this methodology to other events.

2.5 Facility Popularity Assessment from Fine Grained Twitter Analysis

This section expands on facility detection discussed in the previous section, and explores ways to assess facility popularity with Twitter data using natural language processing and text classification methods. Land use data is often a crucial variable in geographic research. Population distribution, biodiversity monitoring, urban dynamics, and energy consumption are a few areas that have recently increased in importance, all which land use data is central to. At the moment land use data are often aggregate classes where mixed land uses are aggregated into single high-level category. By disaggregating land use data, we can become more precise and assess unique facility types going from a *commercial* land use class to *restaurant, retail,* and *café* facilities. In combination with the detection of these facility types, assessing their popularity over space and time could provide an enormous amount of insight into population dynamics. Understanding how a facility's popularity changes from the daytime into nighttime, over the course of a week, and during special events could be very beneficial for urban planners and general geographic research. Yet geospatial data has historically been large, static datasets that get updated over the timescale of years. This multi-year timescale has limited the kind of research questions that can be asked and the level of certainty in which predictions can be made. There is a need for the integration of near real-time, dynamic geospatial datasets in geographic research to allow for predictions at a finer temporal scale. With these types of data, we can advance from datasets like Census derived population information which generally tells us where people live, to dynamic near real-time information which can tell us where people are likely to be at various times throughout the day.

With the continued growth of social media, there are more opportunities to use this data for facility detection and popularity assessment. Services like Twitter and Facebook allow users to connect with friends and broadcast thoughts, and have the ability to record temporal and spatial information associated with the users' posts. Some advantages of using social media data is the near real-time nature of it and its ability to provide data on not only where something or someone is, but context for why they are there. Past research has shown the effectiveness of using large scale social media *check-in* data to classify land use (Zhan et al. 2014; Lansley and Longley 2016). When a user *checks-in* to a specific place on social media, they explicitly state that they are at a specific location, at a specific point in time. If monitored over time, the difference in the number of *check-ins* at a location can provide insight into when a specific type of land use in a geographic location becomes popular, and how that popularity changes over time. This record of *checking-in* provides a high degree of confidence in the locational and temporal accuracy of the data but accounts for a relatively small percentage of geotagged social media data. There are still a lot of geotagged non-check-in social media data that is ignored that holds valuable information regarding land use and facility type popularity.

We aim to add to this research by exploring the possibility of removing the need for explicit *check-in* social media data, specifically using Twitter, using only the text associated with the social media post for facility detection and popularity. By incorporating non-explicit *check-in* social media data into the classification process there is an opportunity to significantly increase the amount of data we have access to (thereby increasing the geographic range of locations we could classify) and increase the level of certainty when classifying land use and facility type popularity.

When using non-check-in social media data, the only relevant information left, other than the geotagged latitude longitude coordinates, is the text associated with the social media post. A user might share relevant information about being at a location of interest (e.g. a restaurant or airport) in the text of the post, and not explicitly *check-in* to that place. Yet, the geotagged social media post might still be as useful as the corresponding *check-in* data if we can learn to interpret and understand what is being said in the text of the post. For this we use natural language processing tools and machine learning algorithms to classify unstructured text into whether or not a user is seemingly at a location of interest when the social media post was made. The first tweet in Fig. 2.6 shows an example of when a user is talking about airports but is seemingly not at an airport when the tweet was sent, whereas the second tweet suggests the user is seemingly at an airport when the tweet was sent. We use natural language processing tools to convert Twitter text into a statistical numeric representation of the tweet in the form of Term Frequency–Inverse Document Frequency (TF–IDF) vectors. TF–IDF vectors are then used as an input for machine learning algorithms to predict whether or not the user is seemingly at the location of interest. We compare the performance of Naïve Bayes and Support Vector Machine algorithms in the process of classifying Twitter text. Once we determine if a tweet is relevant, we can access the locational and temporal metadata associated with each tweet in our dataset. Our machines learning classifiers achieve an average accuracy of approximately 85% varying across multiple facility types in correctly identifying tweets where users are seemingly at the location they are talking about in the tweet.

Fig. 2.6 Examples of tweets with a focus on airports. The first tweet is an example of data that might not be useful for facility popularity, whereas the second tweet might be useful and suggests the user is seemingly at an airport when the tweet was sent

2.5.1 Basic Description

For this study, we have focused on collecting Twitter data extracted from the Twitter Streaming API for the whole world. We have deployed a Linux cluster of seven machines, and each machine is responsible for collecting data from one of the seven continents of the world. This is made possible by using Twitter Streaming API's bounding box parameter, which requires south-west and north-east geographic coordinates for the region under consideration. One year of Twitter data (06/15/2015–06/15/2016) and only English language tweets are used in this study for training and classification purposes. We have used PlanetSense architecture to harvest twitter data and the streaming tweets are stored on an ElasticSearch cluster in real-time (Thakur et al. 2015). On an average, we have collected more than 8.5 million tweets per day. Later, Spring.io ES connector is used to pull the *tweeted text* from ElasticSearch cluster. We have written custom Java code to filter-out emoticons and non-ASCII characters before we perform training and classification. The three facility types that we analyzed in this paper are Restaurants, Airports, and Stadiums. Each tweet in the Restaurant dataset has the word *restaurant* in the tweet. This includes variations of the word *restaurant* including upper and lower case (i.e. *Restaurant, restaurant, RESTAURANT*, etc.), plural (i.e. *restaurants*) and hash-tags (i.e. *#restaurant*). The Airports and Stadiums dataset is the same except with the words *airport* and *stadium*. The Restaurants and Stadiums dataset has a total of 2000 unique, English language, geotagged tweets. The Airports dataset has a total of 1600 unique, English language, geotagged tweets.

2.5.2 Twitter Training Data

The 5600 tweets total dataset were hand labeled by the authors as either one of two classes: either [A] a tweet where the user seemingly indicates that they are at a specific facility type (i.e. restaurant, airport, or stadium) at the time the tweet was sent, or [B] talking about the specific facility type abstractly or in the past or future tense, suggesting that the user is not at the facility location when the tweet is sent. For example, the following tweets represent each class described above:

[A]   Just waiting on my food (at Tracks End Restaurant in Chicago, IL)
[B]   H3796 [NEW] Authorizing the city of Northampton to issue five above quota annual all-alcohol restaurant licenses.

The Restaurants and Stadiums dataset has an even 1000 class A tweets and 1000 class B tweets split. The Airports dataset has 800 class A tweets and 800 class B tweets. The original 5600 tweets did not immediately provide equal splits of class A and class B tweets for each facility dataset. More tweets were randomly sampled when needed to ensure equal splits of classes. For example, the original 2000 tweets randomly sampled for the Restaurants dataset, did not exactly provide an even split

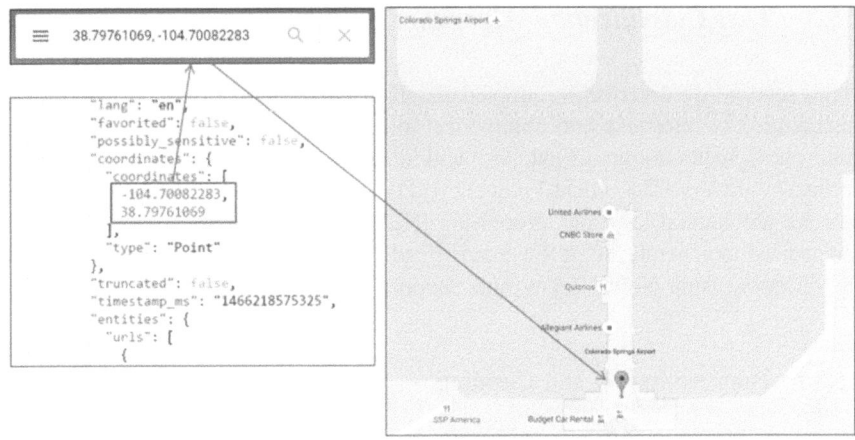

Fig. 2.7 An example of how we check our hand labeling of tweets. We look up the coordinates of the tweet, search for the coordinates, and see if the corresponding location is the place talked about in the tweet

1000 class A tweets and 1000 class B tweets. Once 1000 tweets for either class was reached, no more tweets were included, and extra random samples occurred until we reached 1000 for both classes.

To provide ground truth validation, we looked up the latitude/longitude coordinates for random tweets in class A for Restaurants, Airports, and Stadiums to see what exists at those locations (Fig. 2.7).

For the Restaurants dataset, 46 tweets were located at the location described in the tweet, with 4 being considered incorrect. From this, we can suggest that our hand labeling is approximately 92% accurate for Restaurants in classifying whether the user was at the talked about location in the tweet, when the tweet was sent. For the Airports dataset, 49 tweets were located at the location described in the tweet, with 1 being considered incorrect. This suggests our labeling is approximately 98% accurate for Airports. We checked approximately 10 tweets for Stadiums and all were located at the location described in the tweet. While we did not check 50 class A tweets for Stadiums, we expect similar results seen in Restaurants and Airports.

The final Restaurants and Stadiums dataset includes 2000 geotagged class A or class B tweets, and the Airports dataset includes 1600 geotagged class A or class B tweets with an estimated accuracy for all 3 datasets approximately ranging from 90 to 98%. This is important to note as when we discuss model accuracy results in Sect. 6, the model accuracy should be taken in consideration with the hand labeling accuracy stated here. For example, if the hand labeling accuracy is 90% and the model accuracy is 90%, then overall classification accuracy would be approximately 80%.

2.5.3 Text Classification Algorithm and Approach

In this section, we describe the adopted classification algorithm for transforming the unstructured tweeter text into a numerical input for text classification and explain the overall methodology. First, we will discuss how and why we use Term Frequency–Inverse Document Frequency (TF–IDF) vectors. Second, we'll discuss how we use natural language processing (NLP) tools to turn tweets into data that can be used to generate TF–IDF vectors. Finally, we'll discuss how we perform text classification using Naïve Bayes, and Support Vector Machine (SVM) classifiers.

2.5.3.1 Data Sanitation and Cleaning

Before the unstructured text of the Twitter data is used to create TF–IDF vectors, the data goes through a series of filtering and cleaning steps.

After sampling, hash-tags were removed, and the "@" character was replaced with the word at in each of the tweets across both datasets. For example, the following sentence [A] is changed to the following sentence [B]:

[A] Now that's what I call a #beer. #FamilyDayOut2 @ Cargo Restaurant Bar
[B] Now that's what I call a beer. FamilyDayOut2 at Cargo Restaurant Bar.

The "@" character replacement and the hashtag removal are done for natural language processing reasons that will be described in Sect. 7.3. Later on we perform, TFID calculation that represents the frequency of a given word within a document, weighted by how often that word appears in other documents across the rest of the corpus. This provides a statistical representation for how important a given word is to a given tweet. To compute TF–IDF vectors for each tweet, we go through a series of NLP steps using Stanford's CoreNLP package (Manning et al. 2014). CoreNLP was initially released in 2010 and provides a range of features, including giving the base form of words, identifying parts of speech, indicating sentiment, and more.

2.5.4 Machine Learning Classification

At this point, each Tweet in our corpus has a multi-dimensional TF–IDF vector representing it, and an associating label indicating whether or not the user is seemingly at a Restaurant, Airport, or Stadium when the tweet was sent. Once the data is in this format, this problem becomes a standard machine learning classification problem. We have used Naïve Bayes and Support Vector Machine classifiers for our analysis.

2.5.5 Results and Analysis

In this section, we outline the process of training Naïve Bayes and SVM classifiers to classify TF–IDF vectors for Restaurants, Airports, and Stadiums, present the performance of those models, and evaluate the viability of this approach for land use and facility type popularity classification.

2.5.5.1 Training Classifiers

For each facility type TF–IDF dataset, we randomly split the data with 90% of the data for training and 10% for testing. We did these 100 times for each model, and for each facility type dataset. For example, for a Naïve Bayes classifier for the Restaurants dataset we randomly split the Restaurant dataset 90/10, train a Naïve Bayes classifier, and produce accuracy and precision measures for that classifier, and repeat this process 99 more times finding average accuracy and precision values. Then we do this again for SVM, and then repeat this process for the rest of the facility type datasets.

2.5.5.2 Evaluation Metrics

The metrics we use for the evaluation of the classifiers are accuracy and precision. Accuracy of the classifier is defined by the number of True Positive plus True Negative, divided by total number of samples (or Positives plus Negatives).

$$Accuracy = \frac{TP + TN}{P + N}$$

Precision is defined by True Positive, divided by True Positive plus False Positive.

$$Precision = \frac{TP}{TP + FP}$$

True Positives refer to the positive samples that were correctly labeled by the classifier (i.e. the hand label of the tweet was class **A** and the classifier predicted class **A**). True negatives refer to the negative samples that were correctly labeled by the classifier (i.e. the hand label of the tweet was class **B** and the classifier predicted class **B**). False Positive refers to the negative samples that were incorrectly labeled as positive (i.e. the hand label of the tweet was class **B** and the classifier predicted class **A**).

We are most interested in precision as a metric. For the purposes of using tweets to classify land use and facility popularity, we want to minimize the amount of times the classifier incorrectly predicts a tweet that is of class **B** (the user is not seemingly at the location of interest) as class **A** (the user is seemingly at the location of interest). In this case, the classifier predicts false data as true. Whereas it is more acceptable for the classifier to predict true data (class **A**) as false (class **B**) as this means we simply miss out on potentially good data.

2.5.5.3 Results

After 100 iterations of training and testing classifiers for each dataset, the average accuracy and precision metrics can be seen in Table 2.3. The Airports dataset uniquely is classified better in both accuracy and precision when using Naïve Bayes, whereas both the Restaurants and Stadiums datasets favor SVM for accuracy and precision.

2.5.5.4 Accuracy

SVM accuracy for Restaurants provides approximately a 7% increase, whereas the change in accuracy from Naïve Bayes to SVM for Airports and Stadiums are less significant. Restaurants achieve the highest classification accuracy at 93% with Airports and Stadiums achieving peak accuracies of 85 and 82% respectively (Fig. 2.8). We speculate that Restaurants achieve a higher accuracy due to the expected time spent at restaurants by customers, and by the nature of activity that occurs at restaurants. People tend to go to restaurants for a singular reason and spend anywhere from 1 to 2 h there. For airports and stadiums, people tend to stay at those facilities for extended periods of time, sometimes for long layovers at airports. This allows for a wider variety of things to occur and to be witnessed and tweeted about at airports and stadiums.

The average accuracy distributions for the three facility types show relatively normal distributions with standard deviations ranging from 0.024 to 0.028 with the Restaurant SVM classifier as an exception with a standard deviation of 0.016 (Fig. 2.9).

Table 2.3 Average accuracy and precision values for Naïve Bayes and SVM after 100 iterations

	NB accuracy	SVM accuracy	NB precision	SVM precision
Restaurants	0.850	0.932	0.892	0.987
Airports	0.851	0.824	0.896	0.811
Stadiums	0.804	0.823	0.797	0.829

Fig. 2.8 Average accuracies for Naïve Bayes and SVM

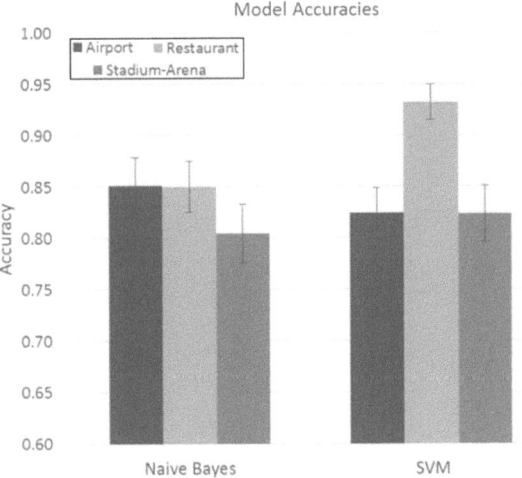

Fig. 2.9 Average accuracy distributions for Naïve Bayes and SVM. The y-axis shows how many times out of the 100 iterations that classifier achieved a given accuracy (x-axis)

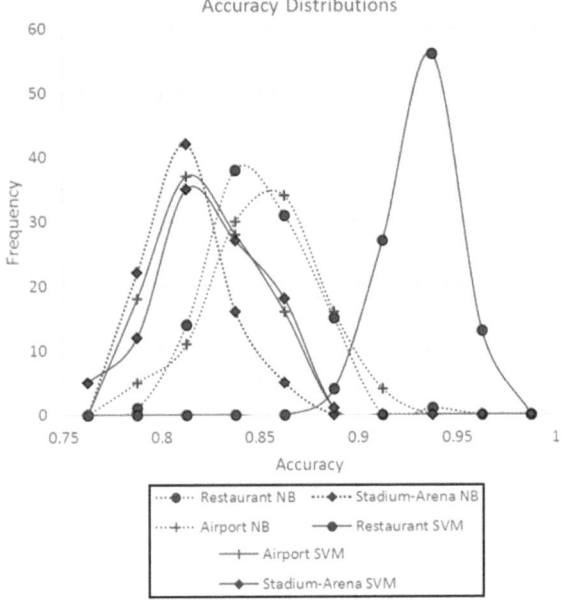

2.5.5.5   Precision

SVM provides the highest precision values for Restaurants and Stadiums with Naïve Bayes providing the highest precision values for Airports (Fig. 2.11). Restaurants achieve the highest average precision of 0.98 with Airports and Stadiums achieving peak average precision of 0.89 and 0.82 respectively.

Fig. 2.10 Average precision
for Naïve Bayes and SVM

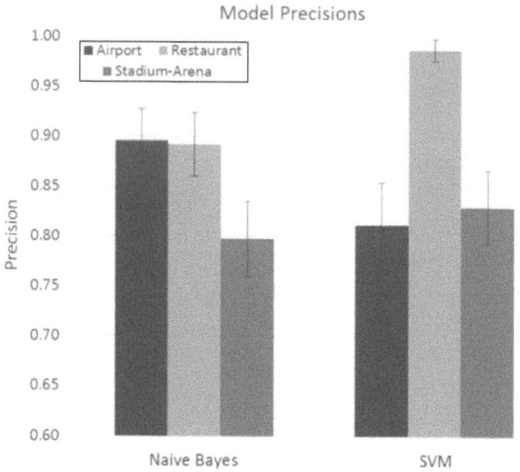

The average precision distribution for the 3 facility types show slightly wider
distributions compared to the average accuracy distributions, with the anomaly of
Restaurants (Fig. 2.10). The SVM classifier for Restaurants consistently achieves
high precision with a standard deviation of 0.011. The rest of the standard deviation
values range from 0.031 to 0.042 (Fig 2.11).

Fig. 2.11 Average precision
distributions for Naïve Bayes
and SVM. The y-axis shows
how many times out of the
100 iterations that classifier
achieved a given precision
(x-axis)

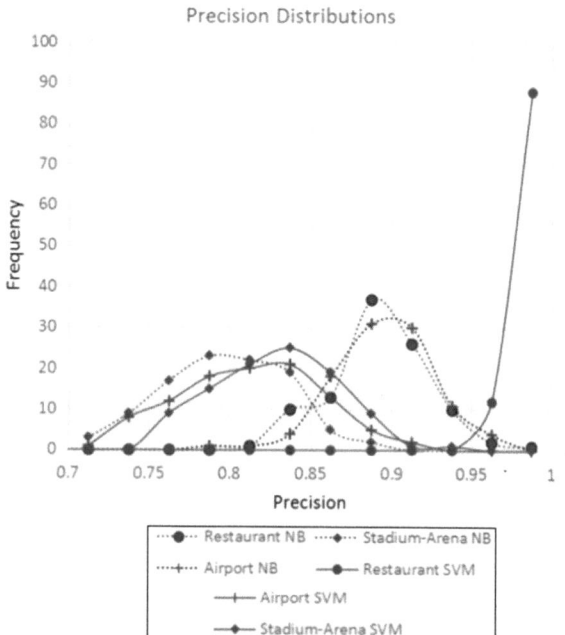

As previously mentioned, precision is the most important evaluation metric for the purposes of our study. It is more important to ensure that the classifier does not predict false data as true. A peak precision value of 0.98 for Restaurants indicates that while the overall accuracy is closer to 93%, the classifier will hardly ever predict false data as true.

Recalling Sect. 7.6.3 describing the hand labeling of tweets to make training data, we evaluated that we were approximately 90–98% accurate when deciding whether a user was seemingly at the location they suggest they are when the tweet was sent. This accuracy needs to be taken in consideration with the classifiers accuracy/precision. If Restaurant precision is 0.98, and we conservatively estimate that our hand labeling accuracy of that data is 90%, then we can assume that approximately 88% of the time our Restaurant classifier predictions are accurate.

2.6 Summary

In this chapter, we presented a need to augment our understanding of human dynamics by using novel data measurements and analytics. We focused our studies to use geo-located human data generated through mobile phones and social media activity patterns. To harvest this data at scale, we proposed a scalable big data architecture that allows us to harvest and analyze real-time geotagged social media data. We presented several case studies that demonstrated the efficacy of using such data for studying human dynamics in changing world. We also presented, how human activity patterns can aid in discovering new facility types and population locations. We present results of using machine learning and natural language processing for text classification on Twitter data for the purposes of classifying human population distribution, land use and facility popularity. While there are some limitations to use this data, for example veracity and completeness, overall the studies presented here make a strong case of using geo-located data to gain insight into human dynamics at a fine resolution, otherwise never been achieved before.

Acknowledgements This manuscript has been authored by UT-Battelle, LLC, under Contract No. DE-AC05-00OR22725 with the U.S. Department of Energy. The United States Government retains, and the publisher, by accepting the article for publication, acknowledges that the United States Government retains a non-exclusive, paid-up, irrevocable, world-wide license to publish or reproduce the published form of this manuscript, or allow others to do so, for United States Government purposes.

References

Becker, H., Naaman, M., & Gravano, L. (2010). Learning similarity metrics for event identification in social media. In *Proceedings of the Third ACM International Conference on Web Search and Data Mining—WSDM '10* (p. 291).

Berry, M. W., Browne, M., Langville, A. N., Pauca, V. P., & Plemmons, R. J. (2007). Algorithms and applications for approximate nonnegative matrix factorization. *Computational Statistics & Data Analysis, 52*(1), 155–173.

Bhaduri, B., Bright, E., Coleman, P., & Urban, M. L. (2007). LandScan USA: A high-resolution geospatial and temporal modeling approach for population distribution and dynamics. *GeoJournal, 69*(1–2), 103–117.

Birkin, M. (2013). The spatial analysis of short-term population movements with social media data. *GeoComputation.*

Bukhari, I., Wojtalewicz, C., Vorvoreanu, M., & Dietz, E. (2012). Social media use for large event management. In *IEEE Conference on Homeland Security Technologies* (no. 1, pp. 24–29).

Charles-Edwards, E., & Bell, M. (2013). Estimating the service population of a large Metropolitan University Campus. *Applied Spatial Analysis and Policy, 6*(3), 209–228.

Cheng, Z., Caverlee, J., & Lee, K. (2010). You are where you tweet. In *Proceedings of the 19th ACM International Conference on Information and Knowledge Management—CIKM '10* (p. 759).

Croitoru, A., Stefanidis, A., Radzikowski, J., Crooks, A., Stahl, J., & Wayant, N. (2012). Towards a collaborative geosocial analysis workbench. In *Proceedings of the 3rd International Conference on Computing for Geospatial Research and Applications—COM.Geo '12* (pp. 18:1–18:9).

Croitoru, A., Crooks, A., Radzikowski, J., & Stefanidis, A. (2013). Geosocial gauge: A system prototype for knowledge discovery from social media. *International Journal of Geographical Information Science, 27*(12), 2483–2508.

Di Gregorio, A., & Jansen, L. J. M. A. (1998). New concept for a land-cover classification system. In *Proceedings of the Earth Observation and Environmental Information 1997 Conference, Alexandria, Egypt* (vol. 2, no. 1, pp. 55–65).

Donnay, J.-P., & Unwin, D. (2001). Modelling geographical distributions in urban areas. *Remote Sensing Urban Analysis*, 205–224.

Elwood, S., Goodchild, M. F., & Sui, D. Z. (2012). Researching volunteered geographic information: Spatial data, geographic research, and new social practice. *Annals of the Association of American Geographers, 102*(3), 571–590.

Frias-Martinez, V., Soto, V., Hohwald, H. & Frias-Martinez, E. (2012). Characterizing urban landscapes using geolocated tweets. In *2012 International Conference on Privacy, Security, Risk and Trust and 2012 International Conference on Social Computing* (pp. 239–248).

González, M. C., Hidalgo, C. A., & Barabási, A.-L. (2008). Understanding individual human mobility patterns. *Nature, 453*(7196), 779–782.

Goodchild, M. F. (2007). Citizens as sensors: The world of volunteered geography. *GeoJournal, 69*(4), 211–221.

Grace, J. H., Zhao, D., & danah boyd. (2010). Microblogging. In *Proceedings of the 28th of the International Conference Extended Abstracts on Human Factors in Computing Systems—CHI EA '10* (p. 4517).

Hudson-Smith, A., Batty, M., Crooks, A., & Milton, R. (2009). Mapping for the masses: Accessing web 2.0 through crowdsourcing. *Social Science Computer Review, 27*(4), 524–538.

Java, A., Song, X., Finin, T., & Tseng, B. (2007). Why we twitter. In *Proceedings of the 9th WebKDD and 1st SNA-KDD 2007 Workshop on Web Mining and Social Network Analysis—WebKDD/SNA-KDD '07* (pp. 56–65).

Jochem, W. C., et al. (2013). Estimating traveler populations at airport and cruise terminals for population distribution and dynamics. *Natural Hazards, 68*(3), 1325–1342.

Kobayashi, T., Medina, R. M., & Cova, T. J. (2011). Visualizing diurnal population change in urban areas for emergency management. *The Professional Geographer, 63*(1), 113–130.

Kuhn, W. (2012). Core concepts of spatial information for transdisciplinary research. *International Journal of Geographical Information Science.* http://dx.doi.org/10.1080/13658816.2012. 722637.

Lansley, G., & Longley, P. A. (2016). The geography of Twitter topics in London. *Computers, Environment and Urban Systems, 58,* 85–96.

Manning, C. D., Bauer, J., Finkel, J., Bethard, S. J., Surdeanu, M., & McClosky, D. (2014). The Stanford CoreNLP natural language processing toolkit. In *Proceedings of 52nd Annual Meeting of the Association for Computational Linguistics: System Demonstrations* (pp. 55–60).

Martin, D., Cockings, S., & Leung, S. (2015). Developing a flexible framework for spatiotemporal population modeling. *Annals of the Association of American Geographers, 105*(4), 754–772.

Meyer, W., & Turner, B. L. (1996). Land-use/land-cover change: Challenges for geographers. *GeoJournal, 39*(3), 237–240.

Morstatter, F., Pfeffer, J., Liu, H., & Carley, K. (2013). Is the sample good enough? Comparing data from Twitter's streaming API with Twitter's firehose. In *Proceedings of ICWSM* (pp. 400–408).

Reades, J., Calabrese, F., & Ratti, C. (2009). Eigenplaces: Analysing cities using the space–time structure of the mobile phone network. *Environment and Planning B: Planning and Design, 36* (5), 824–836.

Seung, H. S., & Lee, D. D. (1999). Learning the parts of objects by non-negative matrix factorization. *Nature, 401*(6755), 788–791.

Stewart, R. et al. (2017). *Can social media play a role in the development of building occupancy curves?* (pp. 59–66). Cham: Springer.

Thakur, G. S., Bhaduri, B. L., Piburn, J. O., Sims, K. M., Stewart, R. N., & Urban, M. L. (2015). PlanetSense: A real-time streaming and spatio-temporal analytics platform for gathering geo-spatial intelligence from open source data. In *Proceedings of the 23rd SIGSPATIAL International Conference on Advances in Geographic Information Systems* (vol. 03–06– November, pp. 11–14).

Turner, A. (2006). *Introduction to neogeography.* California: O'Reilly.

Zhan, X., Ukkusuri, S. V., & Zhu, F. (2014). Inferring urban land use using large-scale social media check-in data. *Networks and Spatial Economics, 14*(3–4), 647–667.

Author Biographies

Gautam Thakur is a research scientist at Oak Ridge National Laboratory. He received his Ph.D. degree in computer information and science engineering from University of Florida in 2012. His research focuses on population dynamics, urban morphology, network science, and transportation simulation and modeling.

Kelly Sims is a Post-Master's Research Associate at the Oak Ridge National Laboratory. She currently supports the Department of Energy's GIS spatial research initiatives, with particular emphasis on spatio-temporal population and social dynamics modeling and simulation. She is interested in characterizing the spatiotemporal and demographic patterns of populations to inform a variety of applications including urban resiliency, human health and security management and response. Other interests include data collection, input and quality assurances, as well as multi-resolution data analysis.

Huina Mao received his Ph.D. degree in Informatics from Indiana University in Bloomington in 2014. At the time of co-writing the chapter, she was a Liane Russell Fellows at the Oak Ridge

National Laboratory in the US. Her research interests include big data, natural language processing, machine learning, social science, and urban science.

Jesse Piburn holds an M.S. from University of Tennessee. He is a Research Scientist in Geographic Data Sciences at the Oak Ridge National Laboratory. His research interests and experience include scalable geographic data science, spatiotemporal analytics, and machine learning as applied to national security, urban dynamics, and clean energy.

Kevin Sparks holds a M.S. from Pennsylvania State University. He is a Research Associate interested in Geographic Information Science and Technology at Oak Ridge National Laboratory. His research interests include improving our understanding of the fundamental cultural and regional effects that shape the dynamics of human societies and cities at an aggregate scale through the use of open geosocial data and spatiotemporal analysis.

Marie Urban received her M.S. from Purdue University. She is a Research Scientist in the Population Distribution and Dynamics Team, Geographic Information Science and Technology Group at the Oak Ridge National Laboratory. Her current research interests consist of open source research including social media in support of higher temporal and spatial resolution modeling and visualization to capture population dynamics through building use.

Robert Stewart received his Ph.D. from University of Tennessee. He is a geographic data scientist in the Geographic Information Science and Technology group at the Oak Ridge National Laboratory and an adjunct assistant professor in Geography at the University of Tennessee. His work is focused on statistical and computational methods in the areas of spatial and spatiotemporal modeling, with an emphasis on uncertainty, risk, and decision analytics. Areas of application include population dynamics, sociocultural/economic analytics, social media, and environmental risk. Quantifying uncertainty and risk as well as understanding the implications of both for decision making is a recurring theme in his work.

Eric Weber received his M.A. from University of Kansas. He is a Research Scientist at the Oak Ridge National Laboratory. His research is focused on improving spatial and temporal resolution in the mapping of human populations by integrating diverse data sources including census and survey data, volunteered geographic information, social media data, and high resolution remote sensing-derived settlement and building data.

Budhendra Bhaduri received his Ph.D. from Purdue university. He is a Corporate Research Fellow and Director, Urban Dynamics Institute, and Group Leader, Geographic Information Science and Technology at the Oak Ridge National Laboratory. His research interests and experience include novel implementation of geospatial science and technology, namely, integration of population dynamics, geographic data science, and scalable geocomputation for addressing high resolution, spatially explicit modeling and simulation of complex urban systems namely, the critical intersection of energy, human dynamics, and urban sustainability.

Chapter 3
Uncovering the Relationships Between Phone Communication Activities and Spatiotemporal Distribution of Mobile Phone Users

Yang Xu, Shih-Lung Shaw, Feng Lu, Jie Chen and Qingquan Li

3.1 Introduction

The pulses of our cities are largely driven by human activities and their movements. An improved understanding of where people are in space and time would benefit urban and transport planning, and facilitate academic research in a wide range of disciplines (e.g., geography, epidemiology, and economics). Traditionally, our abilities to capture spatial and temporal patterns of population distributions largely rely on census data. Despite of their usefulness in population studies, the collection

Y. Xu
Department of Land Surveying and Geo-Informatics, The Hong Kong Polytechnic University, Kowloon, Hong Kong
e-mail: yang.ls.xu@polyu.edu.hk

S.-L. Shaw (✉)
Department of Geography, University of Tennessee, Knoxville, TN 37996, USA
e-mail: sshaw@utk.edu

S.-L. Shaw
Guangzhou Institute of Geography, 100 Xianlie Zhong Road, Guangzhou, Guangdong 510070, People's Republic of China

F. Lu · J. Chen
State Key Laboratory of Resources and Environmental Information System, Institute of Geographic Sciences and Natural Resources Research, Chinese Academy of Sciences, Beijing 100101, People's Republic of China
e-mail: luf@lreis.ac.cn

J. Chen
e-mail: chenj@lreis.ac.cn

Q. Li
Shenzhen Key Laboratory of Spatial Smart Sensing and Services, Shenzhen University, Shenzhen 518060, People's Republic of China
e-mail: liqq@szu.edu.cn

© Springer International Publishing AG, part of Springer Nature 2018
S.-L. Shaw and D. Sui (eds.), *Human Dynamics Research in Smart and Connected Communities*, Human Dynamics in Smart Cities,
https://doi.org/10.1007/978-3-319-73247-3_3

of census data is costly and time consuming. Moreover, such data provide a static view of population estimates, with update cycles that are relatively long (e.g., 10 years). These issues limit the usability of census data in many application domains, especially the ones (e.g., traffic management, disaster response, and epidemic control) that require timely and spatially detailed population information. Although the ways of estimating population distributions have been enhanced in the past few decades (Dobson et al. 2000; Harvey 2002a, b; Balk 2004; Bhaduri et al. 2007; Stevens et al. 2015), we are still in need of cost-effective ways to capture the whereabouts of people in space and time, which are highly dynamic in its nature.

In recent years, mobile phone data have received much attention in geography and other fields. Several advantages make mobile phone data a valuable resource for studying population dynamics: (1) a high and growing penetration rate of mobile phones around the world[1], (2) various location-aware technologies used in mobile phone positioning (Birenboim and Shoval 2015), and (3) ease of data collection (e.g., little burden on individual participants). Two types of mobile phone data, Erlang and call detail records (CDRs), have been widely used in existing literature to study population distributions (Ahas et al. 2007; Girardin et al. 2009; Reades et al. 2009; Sevtsuk and Ratti 2010). These studies regard phone communication activities as an indicator of the presence of urban population. However, these mobile phone data reveal partial aspects of population dynamics, given the fact that Erlang measures aggregate call volume at cellphone towers, and CDRs are generated during particular types of cellphone activities (i.e., initiating or receiving a phone call/text message). It means many previous studies implicitly assume that phone communication activities could properly reflect the distribution of urban population. Nevertheless, whether this assumption holds has not been investigated. Moreover, few studies have even examined whether phone communication activities could reflect the spatiotemporal distribution of mobile phone users.

To fill the research gap, this study uses a mobile phone data set collected in Shanghai, China to answer an important research question: to what extent could phone communication activities reflect the spatiotemporal distribution of mobile phone users? The mobile phone data set used in this study consists of CDRs plus other cellphone-related logs such as cellular handover and periodic location update. To answer the research question, we extract all CDRs into a separate data set to capture the intensity of mobile phone communications at different places in the city over time. Meanwhile, the complete data set is used to derive the spatiotemporal distribution of mobile phone users. Then, correlation and regression analyses are performed to evaluate the relationships between the two types of distributions. The research findings could reveal the potential bias of using phone communication intensity to reflect the underlying population distribution, and provide useful

---

[1]According to the International Telecommunication Union (ITU 2015), there are more than 7 billion mobile phone subscriptions by the end of 2015, corresponding to a penetration rate of 97%. The penetration rate in developed countries reaches 121% by the end of 2014 (World Telecommunication Development Conference 2014).

information and guidelines of using large-scale mobile phone data in urban dynamics research.

3.2 Related Work

The advent of mobile phones has changed how people interact with the outside world (Schwanen and Kwan 2008). It also transforms the ways human activities are sensed and understood. Mobile phone location data, which suggest locations visited by people, have been used to better understand different aspects of human dynamics. For example, there have been many studies which use Erlang data to examine the rhythms of urban mobility patterns (Ratti et al. 2006; Reades et al. 2009; Sevtsuk and Ratti 2010). In these studies, the intensity of people's phone communication activities is used as an indicator of the presence of urban population. Similarly, call detail records (CDRs) have been used to uncover collective human activity patterns (Candia et al. 2008) and aggregate population movements (Ahas et al. 2007, 2010a). Although it is reasonable to assume a certain degree of correlation between the cellphone usage and the underlying population, the extent to which they are correlated and how their relationships change over space and time need to be further examined and validated. Some studies based on CDRs have used spatiotemporal patterns of cellphone usage (e.g., call volume) to predict land use types (Soto and Frías-Martínez 2011; Pei et al. 2014) and dense urban areas (Vieira et al. 2010). The reliability of these predictions also depends on the assumption of the relationship between cellphone usage and population distribution.

People organize their daily tasks (e.g., sending emails and browsing websites) "on a timescale that is appropriate to its urgency" (Ball 2010, p. 692). Researchers find that individual cellphone usage possesses a "bursty" nature (Candia 2008; Barabási 2010). People could make several phone calls in a short period of time and then none for hours. That means mobile phone data (e.g., Erlang and CDRs) could lead to a biased view of human activities. In recent years, several studies have investigated the bias of mobile phone data in geographical research (Ranjan et al. 2012; Zhao et al. 2016). However, these studies mainly focus on particular aspects of human mobility patterns (e.g., radius of gyration and movement entropy). The relationships between aggregate cellphone usage and population distribution are not examined.

People often spend a large amount of time at specific locations such as home and workplace. Studies have found that mobile phone data can be used to estimate people's activity "anchor" points (Gonzalez et al. 2008; Cho et al. 2011; Xu et al. 2015, 2016). These activity "anchor" points, especially home locations, are used to estimate urban population distributions (Ahas et al. 2010b; Silm and Ahas 2010). However, the anchor-point based approach associates individuals to one or few fixed locations. It thus provides population estimates that are static or at a coarse temporal resolution. Instead, Kang et al. (2012) compares people's cellphone usage and population distributions derived from LandScan data for Harbin, China at a finer temporal resolution (1 h). The authors conclude that the proportion between active mobile subscribers and the actual total population varies in different areas, thus cannot reflect the underlying population properly. However, by using two CDR data sets collected in Portugal and France, Deville et al. (2014) find that the density of active mobile phone users can be used to produce spatially and temporally explicit estimations of population densities at national scales. It appears that researchers have not reached a consensus. It is thus important to look deeper into this issue, which has broad implications for human geography and other related fields.

3.3   Study Area and Mobile Phone Data Set

Shanghai is a century old metropolis. The city has a resident population of 24 million as of 2014 and covers an area of 6340 km^2 (Shanghai Bureau of Statistics 2014). It is the largest city in China by population. As a global financial center, its annual gross domestic product (GDP) was ranked the first among all cities in China in the past five years. The city consists of sixteen administrative districts and the Chongming county (Fig. 3.1a). Eight of them on the west bank of Huangpu River (i.e., Putuo, Zhabei, Hongkou, Yangpu, Jingan, Changning, Xuhui and Huangpu), also known as Puxi, are considered as the historic and commercial center of Shanghai (Fig. 3.1b).

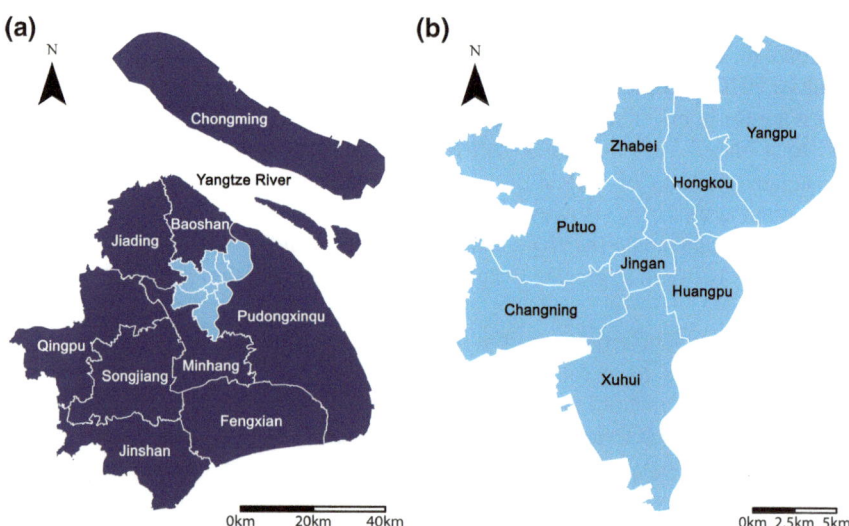

Fig. 3.1 Study area: **a** administrative districts of Shanghai, **b** inset map of the central part (i.e. Puxi) of Shanghai

The mobile phone data set used in this study was collected on a workday in 2012 by a phone service provider in China. As shown in Table 3.1, this data set contains CDRs plus other cellphone-related logs (e.g., regular update, periodic update, cellular handover, power on, and power off). These cellphone-related logs enable us to capture distributions of mobile phone users over space and time better than CDR data. In this data set, each mobile phone record contains information such as the type of event, time (i.e., when the event occurred), and geographic coordinates of the serving cellphone tower. The average nearest distance among cellphone towers operated by this phone service provider in Shanghai is 0.21 km.

Note that we removed mobile subscribers who had *power on* or *power off* event during the study period, since it is difficult to infer their locations when mobile phones are disconnected from the cellular network. The remaining data set after filtering these individuals consists of 698,661 mobile subscribers. As illustrated in Fig. 3.2, we first derive the spatiotemporal distribution of mobile phone users from the complete data set. Meanwhile, we extract all CDRs into a separate data set to capture the intensity of mobile phone communication at different places over time. The relationships between the two types of distributions are then evaluated through correlation and regression analyses.

Table 3.1 Summary of events captured in the mobile phone data set

Type	Event	Description
OT	Phone communication (outbound)	A subscriber makes a phone call or sends a text message
IN	Phone communication (inbound)	A subscriber receives a phone call or text message
RU	Regular update	Triggered by moving from the service area of a cellphone tower to that of another
PU	Periodic update	Triggered by tower pinging if a subscriber has been silent (i.e., no other events detected) for a certain period of time. However, the duration of silence that triggers periodic update is irregular. In addition, mobile phones which are turned off or disconnected from the cellular network do not receive pinging signals from the cellular network
CH	Cellular handover	Transfer of an ongoing phone call from one cellphone tower to another due to a subscriber's movements
ON	Power on	Mobile phone is turned on and connected to the cellular network
OFF	Power off	Mobile phone is turned off and disconnected from the cellular network

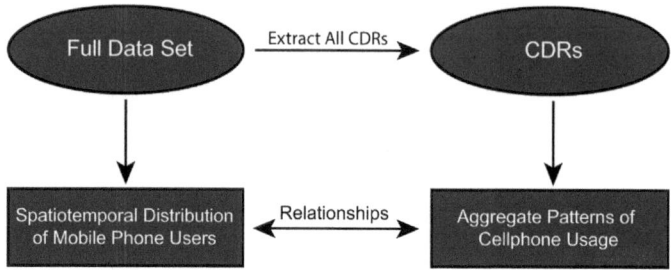

Fig. 3.2 The mobile phone data set consists of call detail records (CDRs) and other cellphone-related logs (e.g., regular update, periodic update, and cellular handover), which make it possible to examine the spatiotemporal relationships between aggregate cellphone usage and phone user distributions

3.4 Research Design

3.4.1 Defining Indicators of Aggregate Cellphone Usage

Mobile phones have become an essential part of people's everyday lives. In recent years, how people use their mobile phones and its societal implications have attracted increasing research interests. When analyzing mobile phone data, previous studies (e.g., Candia et al. 2008; Kang et al. 2012; Yuan et al. 2012) often processed cellphone usage data to reflect either individual phone communication characteristics (e.g., phone call frequency, inter-event time) or collective phone communication activity patterns (e.g., Erlang, call volume). In this study, two indicators of aggregate cellphone usage are selected for the correlation analysis:

– V: *volume of calls/text messages*
– N: *number of active mobile phone users.*

Note that an individual's cellphone trajectory S can be represented as:

$$S = \{P_1(x_1, y_1, t_1, e_1), P_2(x_2, y_2, t_2, e_2), \ldots, P_i(x_i, y_i, t_i, e_i)\} \quad (3.1)$$

where P_i denotes the ith cellphone record; x_i and y_i denote the longitude and latitude of the serving cellphone tower; t_i and e_i represents the time and type of the corresponding mobile phone event (see Table 3.1), respectively.

Given a geographic area A and a time interval T, we define: (1) V_A^T as the total number of phone calls/text messages that occurred within the area A during a time interval T, and (2) N_A^T as the total number of mobile phone users who have made or received at least one phone call/text message within the area A during a time interval T. The two indicators reflect important characteristics of aggregate cellphone usage, and are generated only using the CDRs extracted from the full data set (i.e., records with event type e being IN or OT in Table 3.1).

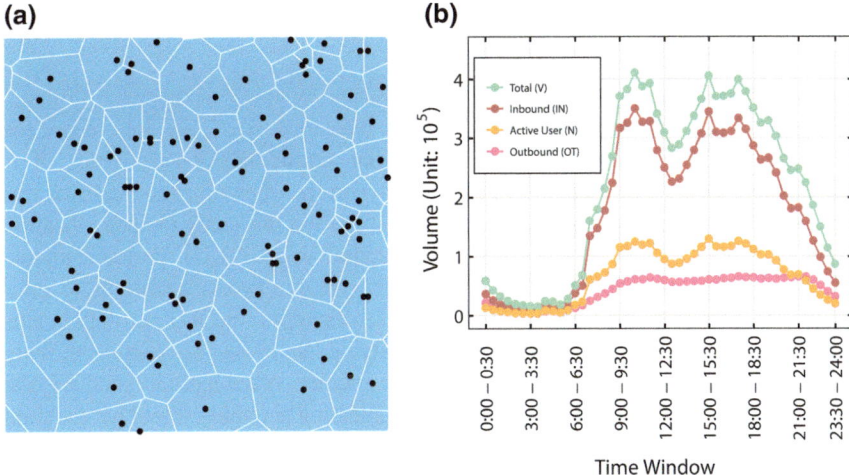

Fig. 3.3 **a** Thiessen polygons are generated based on the spatial distribution of cellphone towers to approximate their service areas. The two indicators (V and N) are calculated at each Thiessen polygon using the mobile phone records that occurred at the corresponding cellphone tower; **b** Global temporal patterns of aggregate cellphone usage at a 30-min time interval (V total volume of phone calls/text messages; N total number of *active* mobile phone users; IN total inbound phone communication activities; OT total outbound phone communication activities)

This study uses Thiessen polygons as the spatial units to derive the cellphone usage indicators. Specifically, Thiessen polygons, which are generated based on the spatial distribution of the cellphone towers, are used to approximate their service areas (Fig. 3.3a). The two indicators can be calculated for each Thiessen polygon using mobile phone records that occurred at the corresponding cellphone tower. Figure 3.3b illustrates the global temporal patterns of the two cellphone usage indicators (as well as the inbound and outbound phone communication activities) at a 30-min time interval. The total volume of phone calls/text messages (V) stays relatively low between midnight and 6:00. It starts to increase in the morning, followed by a fluctuation stage (i.e., 10:00–17:00), and then decreases in the evening. The number of *active* mobile phone users (N) follows a similar pattern of V but has lower intensities. The temporal variations of V and N indicate that the relationship between aggregate cellphone usage and the *total* number of mobile phone users in the city varies greatly throughout the day. However, how their relationships change over space and time remains unclear and is worth an investigation.

3.4.2 Deriving the Spatiotemporal Distribution of Mobile Phone Users

People don't use their mobile phones regularly over time. As CDRs only record the locations visited by people during their phone communication activities, it is questionable to use such data to infer human dynamics when no phone calls or text messages take place. The mobile phone data set used in this study includes location records generated by other events such as regular update (RU), periodic update (PU), and cellular handover (CH). These mobile phone events enable us to infer individual locations at a finer time interval. For example, the RU and CH events allow an individual's location to be continuously updated when he or she is moving from the service area of one cellphone tower to another. When an individual stays at one particular location or has no phone communication activities, his or her location is still reported by the PU event. Thus, the complete data set (i.e., cellphone-related logs along with the CDRs) enables us to estimate a phone user's location at any given time point no matter he or she is moving.

Hence, given an individual cellphone trajectory S, the phone user's location at a particular time point t can be reasonably estimated using the following criteria: (1) if trajectory S contains at least one mobile phone record after time point t, then the phone user's location is estimated as (x_i, y_i) using the mobile phone record $P_i(x_i, y_i, t_i, e_i)$. Here P_i denotes the first mobile phone record which occurred after time point t; (2) if trajectory S has no mobile phone records after t, the mobile phone's location is estimated using the last mobile phone record which occurred before time point t. By doing so, we can estimate each phone user's location at any given time point t, and aggregate all users at the level of cellphone tower service area. These estimates of the spatiotemporal distributions of phone users can be combined with the two cellphone usage indicators for correlation and regression analysis.

It is necessary to note that these estimates are not without uncertainties. On one hand, it is very difficult to pinpoint a mobile phone user's location when it is travelling among different cellphone tower service areas. On the other hand, given the issues of cellphone load balancing or "ping-pong effect" (Isaacman et al. 2012; Csáji et al. 2013), the x, y coordinates of a cellphone tower associated with a particular mobile phone record might not reflect where a user actually stayed. Hence, it is more appropriate to conduct the correlation analysis at a coarser spatial granularity—for example, using a regular grid with a coarser spatial resolution than the cellphone tower service areas—to mitigate the impact of spatial uncertainty.

3.4.3 Correlation and Regression Analysis

Many existing studies, which use mobile phone data for urban mobility research, have an implicit assumption that phone communication activities are highly

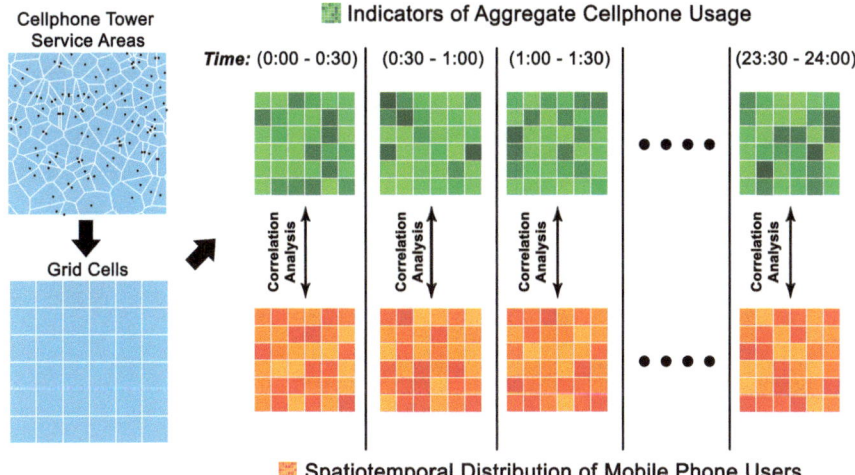

Fig. 3.4 Correlation analysis in a space-time context

correlated to the population size, or at least the number of mobile phone users. While this assumption might hold true, it is important to examine the role of time in such relationships. For example, given a geographic area A, although the number of phone calls/text messages may be roughly the same in early morning (e.g., 07:00–07:30), in late afternoon (e.g., 17:00–17:30), and around midnight (e.g., 23:00–23:30), the total number of mobile phone users observed in each time interval could be quite different from each other. This study conducts a correlation analysis using time as a control factor. As illustrated in Fig. 3.4, we first divide the study area into 1km * 1km regular grid cells. By partitioning a day into forty-eight 30-min time windows, we capture the snapshots of aggregate cellphone usage and the number of mobile phone users in each grid cell for each 30-min time window. These snapshots are used to examine their correlations at different times in a day. We choose the 1km * 1km regular grid in order to obtain the estimates of phone user distribution at a relatively fine spatial resolution while minimizing the spatial uncertainty of mobile phone records.

To perform analysis at the selected spatiotemporal resolution, we first calculate the two indicators of aggregate cellphone usage and the total number of mobile phone users at the level of cellphone tower service areas (i.e., Thiessen polygons). If a mobile phone user has more than one mobile phone record during a particular time window, we use the Thiessen polygon that contains the first mobile phone record as his or her representative location. If there is no mobile phone record during a particular time window, it means this user did not move. Thus, the phone user's location can be estimated using the approach described in the previous section. Once this step is completed, we transform the results onto the grid cells. Considering that a Thiessen polygon could overlap with multiple adjacent grid cells, we clip each Thiessen polygon into sub units. For each sub unit, the indicators of aggregate cellphone usage (V and N) and the total number of mobile phone users (*Pop*) are prorated based on the proportion of its area to the total area of the corresponding Thiessen polygon. We then calculate V, N and *Pop* of each grid cell by adding the values of all sub units that fall within the particular grid cell.

For each time interval T, we first analyze the correlation between the number of mobile phone users (*Pop*) and each of the two cellphone usage indicators using Pearson's correlation coefficients:

$$\rho_{Pop,V}^{T} = \frac{cov(Pop^{T}, V^{T})}{\sigma_{Pop^{T}} * \sigma_{V^{T}}} \tag{3.2}$$

$$\rho_{Pop,N}^{T} = \frac{cov(Pop^{T}, N^{T})}{\sigma_{Pop^{T}} * \sigma_{N^{T}}} \tag{3.3}$$

where: (1) $cov()$ stands for the covariance and σ_X denotes the standard deviation of X; (2) $Pop^{T} = [Pop_{1}^{T}, Pop_{2}^{T}, \ldots, Pop_{m}^{T}]$, $V^{T} = [V_{1}^{T}, V_{2}^{T}, \ldots, V_{m}^{T}]$, and $N^{T} = [N_{1}^{T}, N_{2}^{T}, \ldots, N_{m}^{T}]$; (3) m denotes the total number of grid cells in the study area. The values of $\rho_{Pop,V}^{T}$ and $\rho_{Pop,N}^{T}$ enable us to better assess their correlations during different time periods of a day.

In this study, we introduce two types of regression models that have been suggested in previous studies (Kang et al. 2012; Deville et al. 2014) to further investigate the relationships between the number of mobile phone users and the aggregate cellphone usage:

$$\text{Model 1}: \quad Pop^{T} = a * V^{T} + b \tag{3.4}$$

$$\text{Model 2}: \quad Pop^{T} = a * N^{T} + b \tag{3.5}$$

$$\text{Model 3}: \quad \log_{10}(Pop^{T}) = a * \log_{10}(V^{T}) + b \tag{3.6}$$

$$\text{Model 4}: \quad \log_{10}(Pop^{T}) = a * \log_{10}(N^{T}) + b \tag{3.7}$$

In these regression models, the dependent variable is the total number of mobile phone users in each grid cell during a particular time window (Pop^{T}), and the independent variable is the cellphone usage indicator (V^{T} or N^{T}). Model 1 and Model 2 assume a linear relationship between Pop^{T} and V^{T} (or N^{T}), while Model 3 and Model 4 (i.e., log-transformation models) quantify the power-law relationships between *Pop* and each of the two cellphone usage indicators. The ordinary least squares (OLS) method is used to derive the parameters of these regression models. As the study day is partitioned into forty-eight 30-min time windows, each model produces 48 sets of parameters. We then use three measures, which are the adjusted R^2, the root mean square error (RMSE), and the mean absolute percentage error

(MAPE), to compare the performance of these regression models at different times in a day[2].

3.5 Results and Discussion

3.5.1 Correlation Between the Number of Phone Users and the Two Cellphone Usage Indicators

Figure 3.5a shows the values of $\rho^T_{Pop,V}$ and $\rho^T_{Pop,N}$ and how they change over time. In general, there is a high correlation between the *total* number of mobile phone users (*Pop*) and each of the two cellphone usage indicators during the day time and in the evening. Also, the correlation of *Pop* and the number of *active* mobile phone users (*N*) is always higher than that of *Pop* and the volume of calls/messages (*V*) in the same time window.

According to the temporal variations of $\rho^T_{Pop,V}$ and $\rho^T_{Pop,N}$, the study day can be categorized into several stages. From 07:00 to 21:30, the values of $\rho^T_{Pop,V}$ and $\rho^T_{Pop,N}$ stay above 0.9 and remain relatively stable. A decrease of $\rho^T_{Pop,V}$ and $\rho^T_{Pop,N}$ is observed during 00:00–04:00 and 21:30–24:00, which refer to the time when people have fewer phone communication activities (see Fig. 3.3b). To our surprise, there are some fluctuations of $\rho^T_{Pop,V}$ and $\rho^T_{Pop,N}$ during 03:30–05:00, which is followed by a rising stage (05:00–07:00). These fluctuations, which are somewhat counter-intuitive, encourage us to explore potential explanations. Specifically, we derive several cellphone usage indicators to distinguish inbound and outbound phone communication activities, and further examine their correlations with the number of mobile phones:

- **N_Inbound** Number of active mobile phone users derived from inbound phone communications (*IN*) only
- **N_Outbound** Number of active mobile phone users derived from outbound phone communications (*OT*) only
- **V_Inbound** Volume of inbound calls/messages
- **V_Outbound** Volume of outbound calls/messages.

We find that the correlation coefficients of *Pop* versus *N_Outbound* (i.e., $\rho^T_{Pop,N\_Outbound}$ shown in Fig. 3.5b) and *Pop* versus *V_Outbound* (i.e., $\rho^T_{Pop,V\_Outbound}$ shown in Fig. 3.5c) exhibit smooth temporal variations before 07:00. However, the temporal patterns of $\rho^T_{Pop,V\_Inbound}$ and $\rho^T_{Pop,N\_Inbound}$ are very

---

[2]For Model 3 and Model 4, the three measures (adjusted R^2, RMSE and MAPE) are calculated after converting $\log_{10}(Pop^T)$, $\log_{10}(V^T)$ and $\log_{10}(N^T)$ to the original scale (i.e., Pop^T, V^T, and N^T, respectively).

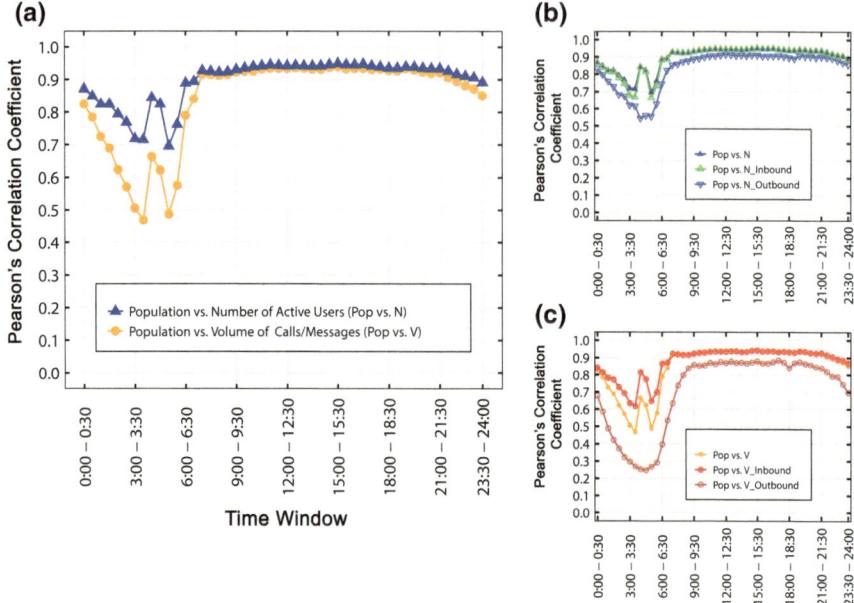

Fig. 3.5 **a** Pearson's correlation coefficients of the *total* number of mobile phone users (*Pop*) and each of the two cellphone usage indicators **b** Pearson's correlation coefficients of *Pop* versus *N_Inbound* (i.e., $\rho^T_{Pop,N\_Inbound}$) and *Pop* versus *N_Outbound* (i.e., $\rho^T_{Pop,N\_Outbound}$). *N_Inbound* denotes the number of *active* mobile phone users derived from the inbound phone communications only. *N_Outbound* denotes the number of *active* mobile phone users derived from the outbound phone communications only; **c** Pearson's correlation coefficients of *Pop* versus *V_Inbound* (i.e., $\rho^T_{Pop,V\_Inbound}$) and *Pop* versus *V_Outbound* (i.e., $\rho^T_{Pop,V\_Outbound}$). *V_Inbound* denotes the volume of inbound calls/text messages, and *V_Outbound* denotes the volume of outbound calls/text messages

similar to that of $\rho^T_{Pop,V}$ and $\rho^T_{Pop,N}$, respectively. This is probably because outbound phone communications are initiated by mobile phone users, while inbound phone communications could include push notifications such as advertisements, weather forecast, news, etc. It is very likely that the fluctuations between 03:30–05:00 are caused by these inbound messages. It also reminds us that the two primary indicators, *N* and *V*, are the combined effects of outbound and inbound phone communications, which not only are related to how people use their mobile phones, but also are related to mobile phones' passive interactions with the outside world.

3.5.2 Comparison of Regression Models

We include four regression models to further examine the relationships between the total number of mobile phone users and each of the two cellphone usage indicators.

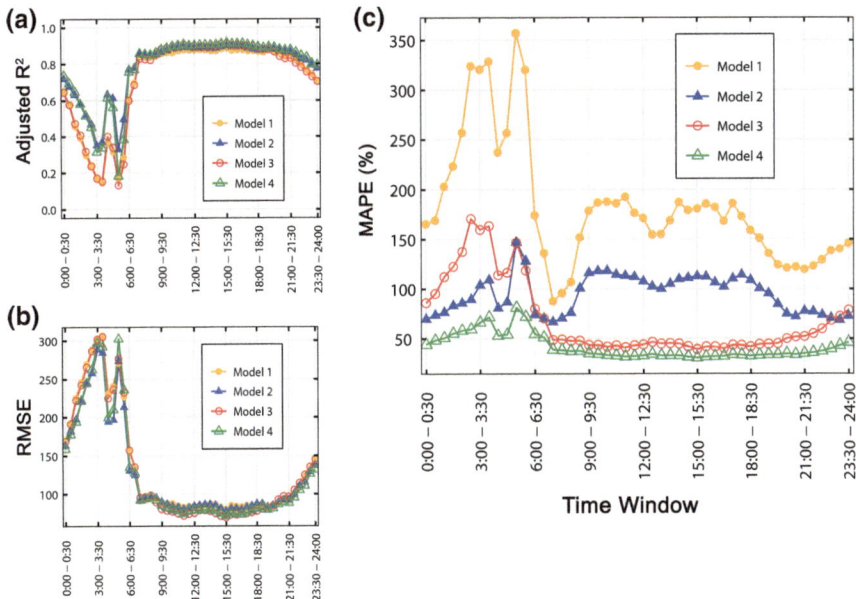

Fig. 3.6 Comparison of the four regression models: **a** Adjusted R^2; **b** Root mean square error (RMSE); **c** Mean absolute percentage error (MAPE)

The adjusted R^2, root mean square error (RMSE), and mean absolute percentage error (MAPE) are used to assess the model performance (Fig. 3.6). As illustrated in Fig. 3.6a, the two models with N^T as the independent variable (i.e., Model 2 and Model 4) have a higher adjusted R^2 than the other two models in each time window. However, when comparing the RMSE of the four models, we find that Model 3 and Model 4 perform better than the other two models during the daytime (07:00–18:00). Notice that the total number of mobile phone users (*Pop*) in the grid cells could vary greatly from each other, it is important to use a normalized measure, which is MAPE in our analysis, to further evaluate the model performance. As illustrated in Fig. 3.6c, Model 3 and Model 4 have a much lower MAPE than the other two models during the daytime. All the three measures suggest that Model 4 performs better than the other three models. The average MAPE of Model 4 between 00:00–07:00 is 58.5%, as compared to 35.1% between 07:00–24:00.

Comparisons of the four regression models indicate that the heterogeneity (or variation) of *Pop* is better explained by the number of *active* mobile phone users (N^T) than by the volume of calls/messages (V^T). As suggested by Barabási (2010), human activities are not random but "bursty"[3]. At a given place during a given time period, V^T is more affected by the individual "burst" of phone communications than

---

[3]In the context of this book, the author refers the word "burst" to brief periods of intensive human activities (e.g., sending text messages) followed by long periods of no activities.

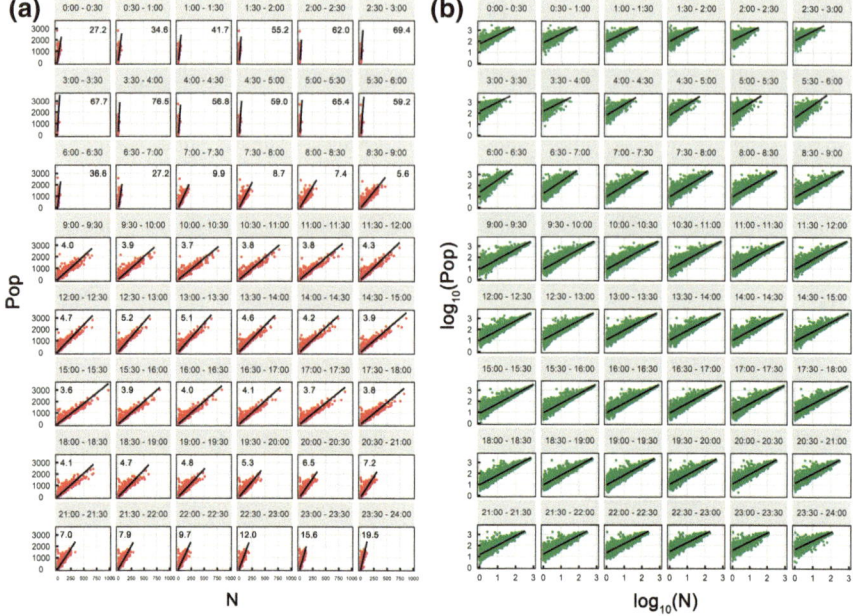

Fig. 3.7 Scatter plots of: **a** *Pop* versus N; **b** $log_{10}(Pop)$ versus $log_{10}(N)$ during different time windows. The black line in each plot denotes the regression line (the numbers in each plot of Fig. 3.7a denotes the coefficient of the regression line)

N^T is, which serves as one potential explanation to our findings. The model comparisons also suggest that the relationships between the total number of mobile phone users and the cellphone usage level are better explained by the log-transformation models (i.e., Model 3 and Model 4) than by the simple linear regression models (i.e., Model 1 and Model 2) when the independent variable (N^T or V^T) is fixed.

Although our findings suggest that the log-transformation models (i.e., Model 4) better describes the relationship between *Pop* and N than the simple linear regression models (i.e. Model 2), it is useful to compare the relationships of *Pop* and N between these two models and examine how the relationships change over time. The scatter plots of *Pop* versus N (Fig. 3.7a) suggest that it is inherently biased to use aggregate cellphone usage to represent the number of mobile phone users. More importantly, the slope of the regression lines indicates that the relationship between *Pop* and N varies greatly throughout a day. That means, even if a place has (or two different places have) the same number of *active* mobile phone users (N) during two different time periods, their values of *Pop* could be quite different from each other.

3.5.3 Cross Validation

We perform a k-fold cross validation to further assess the robustness of Model 4. In particular, the mobile phone data set is partitioned into k subsets with roughly the same size. During the data partition, each individual mobile phone user has an equal probability (i.e., $1/k$) of being assigned to any given subset. Thus, all the subsets after data partition will have approximately the same distribution patterns. During the cross validation process, each time $k - 1$ subsets are used as a *training data set*, and the remaining subset is used as a *validation data set*. The training data set is used to produce the parameters of Model 4, which are then used to predict the total number of mobile phone users (*Pop*) of the validation data set. To ensure that each subset covers an adequate number of samples, we choose $k = 3$ for this particular analysis. Specifically, we perform the 3-fold cross validation 10 times—with each time using a new partition of the mobile phone data set—in order to control the impact of data partition on the analysis results. By doing so, we obtain 30 ($C_3^2 * 10$) pairs of training and validation data sets, and several measures (e.g., average MAPE and average RMSE) of the 30 iterations are used to evaluate the model performance.

Note that we also compare the performance of Model 4 based on: (1) the ordinary least squares (OLS) and (2) the population-weighted least squares (PWLS). The model based on PWLS minimizes the sum of squared residuals weighted by the total number of mobile phone users (*Pop*). Thus, the samples (i.e., grid cells) with smaller *Pop* will have less impact on the regression result. This comparison is expected to generate additional insights into the prediction capability of Model 4.

Figure 3.8 illustrates the model performance of OLS and PWLS. We find that the OLS model generates similar MAPE from the 3-fold cross validation (i.e. green line in Fig. 3.8a) and the full data set (i.e., green line in Fig. 3.6c), which indicates the robustness of Model 4. On the other hand, the OLS model generates lower MAPE (Fig. 3.8a) but higher RMSE (Fig. 3.8b) than the PWLS model. The temporal variations of MAPE/RMSE produced by the two models can be better understood by dividing the study day into two stages:

- **Stage A** refers to the time periods from 00:00 to 07:00 and from 21:30 to 24:00, when the majority of people rest at home. During this stage, people have fewer phone communication activities (Fig. 3.3b). The correlation between *Pop* and the aggregate cellphone usage varies greatly during this stage (Fig. 3.5), which causes notable fluctuations of prediction accuracy (as shown in Fig. 3.8a, b).
- **Stage B** refers to the time period from 07:00 to 21:30, when cellphone communications and other human activities are active. During this stage, the prediction accuracy is much better than that of stage A and remains relatively stable.

We further examine the relationship between the prediction error (i.e. absolute percentage error) and the value of dependent variable (*Pop*) during these two stages

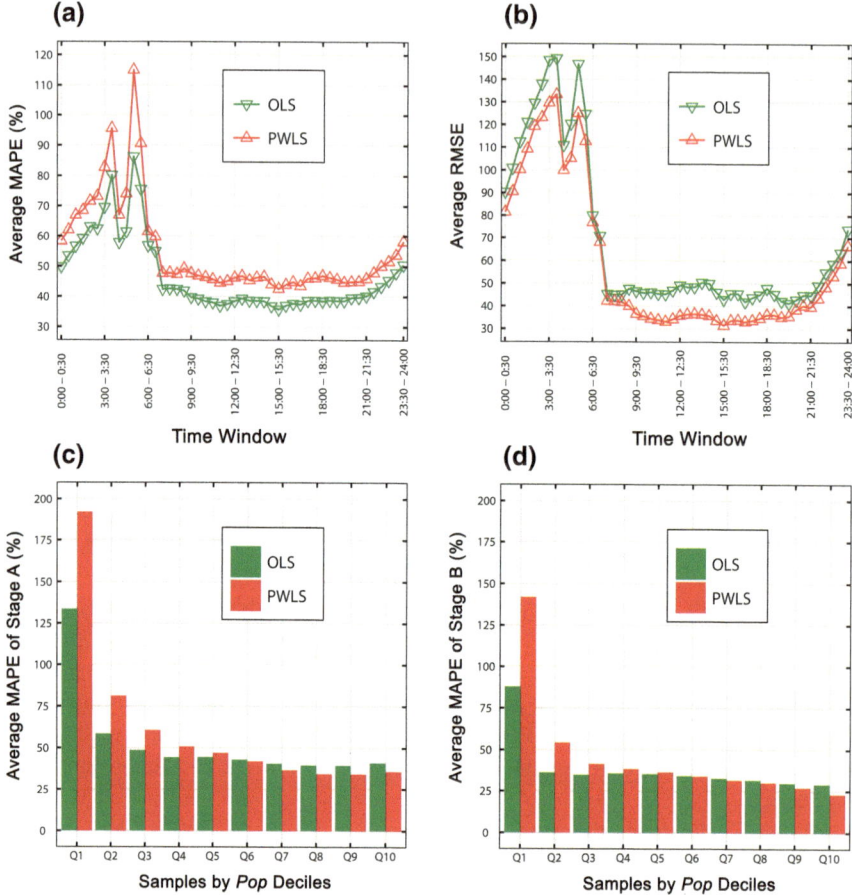

Fig. 3.8 The 3-fold cross validation: **a** Average MAPE of Model 4 based on the ordinary least squares (OLS) method, and the population-weighted least squares (PWLS) method; **b** Average RMSE of OLS and PWLS; **c** Average MAPE of samples by deciles of *Pop* for the stage A (i.e., 00:00–07:00 and 21:30–24:00); **d** Average MAPE of samples by deciles of *Pop* for the stage B (i.e., 07:00–21:30)

produced by OLS and PWLS, respectively. For each model, we aggregate the samples (from validation data sets) during the stage A and the stage B, respectively. For each stage, we organize the samples in an ascending order of *Pop*. We then divide these samples into deciles ($Q1$, $Q2$, … $Q10$) and calculate the average MAPE of each decile. Figure 3.8c illustrates the average MAPE of the samples organized by deciles (generated by the two models) during the stage A. In general, both models yield a better estimation as *Pop* gets larger. Also, PWLS generates better results than the OLS model when the samples have a large value of *Pop* (i.e., $Q6$–$Q10$). An implication is that it is more appropriate to use PWLS than OLS

Table 3.2 The range of value by *Pop* (i.e., the number of mobile phones) decile for stage A (from 00:00 to 07:00 and from 21:30 to 24:00) and stage B (from 07:00–21:30)

Stage A			Stage B		
Pop decile	Minimum value	Maximum value	*Pop* decile	Minimum value	Maximum value
Q1	1.0	20.0	Q1	1.0	8.0
Q2	21.0	32.0	Q2	9.0	13.0
Q3	33.0	45.0	Q3	14.0	18.0
Q4	46.0	61.0	Q4	18.0	25.0
Q5	62.0	80.0	Q5	26.0	34.0
Q6	81.0	109.0	Q6	35.0	46.0
Q7	110.0	150.0	Q7	47.0	63.0
Q8	151.0	214.0	Q8	64.0	92.0
Q9	215.0	304.0	Q9	93.0	168.0
Q10	305.0	984.0	Q10	169.0	1046.0

under certain scenarios (e.g., evacuation) when we want to produce better estimates in populated areas.

Similar patterns are observed for the two models during the stage B (Fig. 3.8d). Both models of this stage, which refers to daytime and early evening, perform relatively well except for Q1 and Q2. The two models, especially PWLS, do not perform well on Q1 and Q2 because these two deciles have a small value of *Pop*. As shown in Table 3.2, the maximum value of *Pop* for Q1 and Q2 during the stage B is 8.0 and 13.0, respectively. The model performance over Q1 and Q2 are affected more by the unique characteristics of individual phone communication activities due to a smaller number of mobile phone users.

The 3-fold cross validation suggests that it is more reasonable to use Model 4 to approximate the relationship of *Pop* and *N*, especially during the daytime and in early evening (i.e. Stage B). After removing the samples with very small values of *Pop* (i.e., Q1 and Q2), we find that the average MAPE of OLS and PWLS models during the stage A changes to 42.6 and 42.7%, respectively. The two models perform better during the stage B, with an average MAPE of 33.0 and 32.8%, respectively. Overall, the PWLS model performs better than the OLS model due to its lower RMSE (Fig. 3.8b).

3.5.4 Spatiotemporal Patterns of Residuals

By performing the 3-fold cross validation, we are able to derive the residuals—measured as the average percentage errors (i.e., $\frac{Y_{predicted} - Y_{observed}}{Y_{observed}} * 100\%$) of 30 iterations using PWLS—at each grid cell for all time windows. A grid cell with a positive or a negative percentage error (during a time window T) suggests that

Model 4 overestimates or underestimates the total number of mobile phone users (*Pop*), respectively. The spatial and temporal patterns of these residuals can help us better understand the relationship between *Pop* and *N*. As Model 4 produces better estimations during the daytime and in early evening (Fig. 3.8), this section only discusses the findings during the stage B, which covers twenty-nine 30-min time windows from 07:00 to 21:30.

The residuals of a grid cell *G* during the stage B can be represented as follows:

$$G = \left\{ erorr^{T_{B1}}, erorr^{T_{B2}}, \ldots, erorr^{T_{Bi}} \ldots, erorr^{T_{B29}} \right\} \tag{3.8}$$

where T_{Bi} denotes the time windows (for example, T_{B1} refers to [07:00–07:30], and TB_{29} refers to [21:00–21:30]). Figure 3.9a shows the residuals of a grid cell that covers part of the *Nanjing Road* pedestrian-only shopping street (Fig. 3.9b), which is one of the world's busiest shopping streets located in Huangpu district in Shanghai. The percentage error of this grid cell remains positive during most of the time windows, which means that the model has mostly overestimated the total number of mobile phone users. The constant overestimation reveals an important fact that a larger percentage of people tend to use their mobile phones in this grid cell (during the daytime and in early evening) as compared to the overall population.

Note that some grid cells may not have mobile phone users (i.e., *Pop* = 0) during particular time windows, which lead to missing values (i.e., $error^{TBi} = NA$) in the residual *G*. As the purpose of this analysis is to find grid cells with similar temporal patterns of residuals, these missing values must be handled appropriately. Figure 3.9c shows the distribution of grid cells with varying number of missing values. It is likely that the grid cells with a large number of missing values reflect less populated areas in Shanghai or areas where mobile phone records are sparse. In this section, we focus on grid cells with no more than 10 missing values (4070 cells in total). As shown in Fig. 3.9d, these grid cells mainly cover the core areas of Shanghai (Fig. 3.1b) and some other administrative districts (e.g., Pudongxinqu, Minhang, Songjiang, Qingpu, Jiading, and Baoshan). As these grid cells have observations during the majority of the time windows, we replace the missing values for each grid cell using a linear interpolation method. For instance, if the percentage error of a grid cell is 10% during time window T_{B1}, and 20% during time window T_{B3}, then the value during T_{B2} (if missing) is estimated as 15%.

We further divide the stage B into three time periods: (1) 07:00–12:00; (2) 12:00–17:00, and (3) 17:00–21:30. For each grid cell, we calculate the percentage of time windows with positive and negative residuals (in each of these three time periods). The temporal patterns of residuals during a particular time period can be characterized as follows:

– If the percentage of time windows with positive residuals is equal to or larger than 75, this time period is labeled as "dominated by overestimations".
– If the percentage of time windows with negative residuals is equal to or larger than 75, this time period is labeled as "dominated by underestimations".
– Otherwise, this time period is labeled as "mixed patterns".

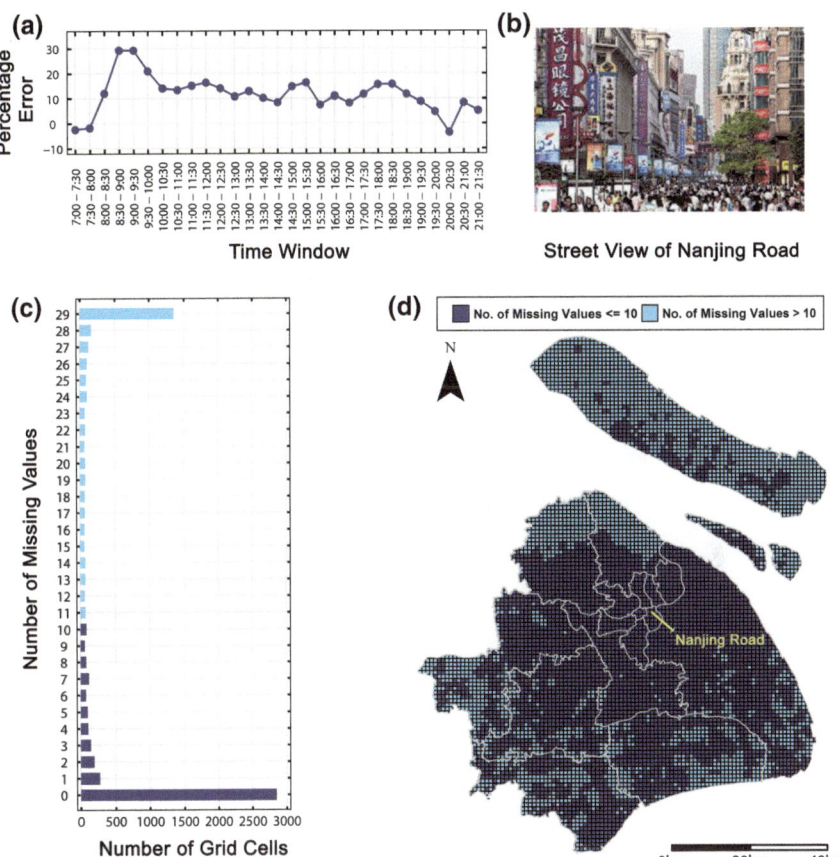

Fig. 3.9 a Temporal variations of residuals (i.e., percentage errors) of a grid cell that covers part of the *Nanjing Road* pedestrian-only shopping street; **b** A street view of *Nanjing Road* (picture from Google Image); **c** Distribution of grid cells with varying number of missing values; **d** Geographic distributions of grid cells with no more than 10 missing values

Figure 3.10 illustrates the major types of grid cells (*C*1–*C*14) with distinct temporal patterns. It is interesting to find that none of these grid cells are mixed with time periods dominated by overestimations (i.e., red segments) and underestimations (i.e., green segments). It is likely that there are some inherent characteristics of the built environment which govern the relationships between the aggregate cellphone usage and the total number of mobile phone users. To better understand the geographic context of these grid cells, we map them onto Google Earth and visually examine some of these places through photos, landmarks, and semantic descriptions. As shown in Fig. 3.11, the grid cells with time period(s) dominated by overestimations (i.e., *C*1–*C*7) cover some important commercial and business areas in Shanghai (i.e., grid cells A to F). At these places, more (i.e., a larger percentage of) people tend to use their mobile phones than the average (percentage) of overall

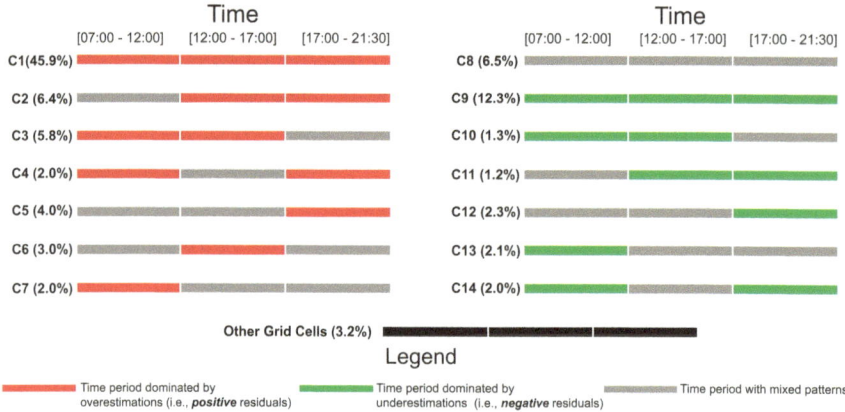

Fig. 3.10 Grid cells with different temporal characteristics (of residuals)

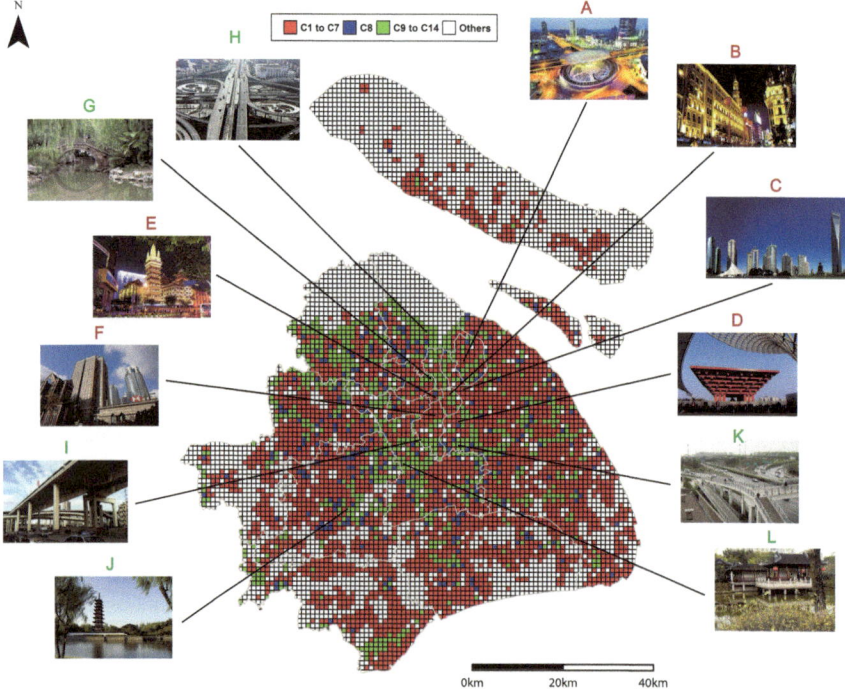

Fig. 3.11 The spatial distribution of grid cells with distinct temporal patterns of residuals. Grid cells A to F, with certain time period(s) dominated by overestimations, refer to some important commercial and business areas in Shanghai. Grid cells G to L, with certain time period(s) dominated by underestimations, represent certain parks (e.g., G, J and L) and places traversed by urban express ways (e.g., H, I, and K), [1]Pictures are captured from Google Image and Panoramio

population. We also find that the grid cells with time period(s) dominated by underestimations (i.e., $C9$–$C14$) include some parks (i.e., grid cells G, J and L) and places traversed by urban express ways (i.e., grid cells H, I and K). At these places, people's cellphone usage is less intense. The temporal patterns of residuals at these selected places suggest that certain characteristics of the built environment—such as land use type, points of interest (POI) and transportation infrastructures—could be considered in the analysis to further understand the behavior of mobile phone usage.

3.6 Conclusion

By using a mobile phone data set that consists of call detail records (CDRs) and other cellphone-related logs (e.g., cellular handover and periodic location update) collected in Shanghai, China, this study evaluates to what extent phone communication activities could reflect the spatiotemporal distribution of mobile phone users. Specifically, we derive two cellphone usage indicators (volume of calls/messages [V] and number of *active* mobile phone users [N]) as well as the *total* number of mobile phone users observed at different places in the city over time, and examine their relationships through correlation and regression analysis. We find that correlations between the number of mobile phone users and each of the two cellphone usage indicators remain high and stable (with Pearson's correlation coefficient above 0.9) during the daytime and in early evening (i.e., 07:00–21:30). Their correlations are generally lower in other time periods, and exhibit notable fluctuations between 00:00–07:00.

We then introduce four regression models (i.e., two simple linear regression models and two log-transformation models) to further examine relationships between the *total* number of mobile phone users (*Pop*) and the two cellphone usage indicators. Several important findings are discovered. First, comparisons of model performance indicate that the number of *active* mobile phone users (*N*) serves as a better independent variable than the volume of calls/messages (*V*) when explaining spatiotemporal distribution of mobile phone users. The volume of calls/messages— at a given place during a particular time period—is likely affected by individual "burst" of phone communication activities (Barabási 2010), which makes the number of *active* mobile phone users (*N*) a better indicator of the mobile phone user distribution. Second, the log-transformation model performs better than the simple linear regression model (in predicting phone user distribution) when the independent variable is fixed. Although the simple linear regression models do not have the best prediction accuracy, our results illustrate that the relationship between the total number of mobile phone users and the cellphone usage level varies greatly throughout a day. It is likely to generate biased results if we use the intensity of aggregate cellphone usage to directly reflect the mobile phone user distribution or the underlying population distribution, and the degree of bias varies with time. Researchers must be cautious when using phone communication activities to quantify certain aspects of urban dynamics. Third, the 3-fold cross validation indicates that the log-transformation model (using V as the independent variable) has a prediction error (i.e., mean absolute percentage error) of 32.8% during the daytime and in early evening (i.e., 07:00–21:30), and 42.7% during other time periods (i.e., 00:00–07:00 and 21:30–24:00). The spatiotemporal patterns of residuals suggest that there exist some inherent characteristics of the built environment which govern the relationships between the cellphone usage and the number of mobile phone users. It suggests that CDR data can be used along with other data sources (e.g., land use type, POI, and transportation infrastructures) to deliver robust estimations of phone user distributions.

Mobile phone data can be leveraged to gain better insights into the whereabouts of people in space and time, which suggests that it serves as a promising data source to supplement traditional approaches (e.g., travel surveys) for studying dynamic population distributions. However, challenges still remain. For example, the mobile phone data used in this study are collected from a single phone company. As a city usually includes multiple phone companies, it is necessary to compare whether the relationships between the cellphone usage level and the distribution of mobile phone subscribers are similar across different cellular networks. How to integrate population estimates from multiple cellular networks in order to gain a more compressive view of urban population distribution is of great importance to applications in emergency response, public health, transport planning, among others.

This research examines only the spatiotemporal relationships between the aggregate cellphone usage and the phone user distributions on a weekday. How their relationships vary between weekdays and weekends, and how such relationships are influenced by special events are not examined in this study. Also, how the spatiotemporal resolutions (e.g., size of grid cell, length of time window) would influence the prediction accuracy is worth a further investigation. Future work can focus on these issues and combine other data sources (e.g., land use type and POI) with CDRs to deliver more robust estimations of mobile phone users and dynamic urban population distributions. Findings of this study provide some useful information and guidelines of using large-scale mobile phone data for geographical studies and urban dynamics research.

Acknowledgements This research was jointly supported by the Alvin and Sally Beaman Professorship and Arts and Sciences Excellence Professorship of the University of Tennessee, Natural Science Foundation of China (41231171, 41371377, 41501486, 91546106, 41571431), Key Program of the Chinese Academy of Science (ZDRW-ZS-2016-6-3), and Beijing Key Laboratory of Urban Spatial Information Engineering (2014101).

References

Ahas, R., Aasa, A., Mark, Ü., Pae, T., & Kull, A. (2007). Seasonal tourism spaces in Estonia: Case study with mobile positioning data. *Tourism Management, 28*(3), 898–910.

Ahas, R., Aasa, A., Silm, S., & Tiru, M. (2010a). Daily rhythms of suburban commuters' movements in the Tallinn metropolitan area: case study with mobile positioning data. *Transportation Research Part C: Emerging Technologies, 18*(1), 45–54.

Ahas, R., Silm, S., Järv, O., Saluveer, E., & Tiru, M. (2010b). Using mobile positioning data to model locations meaningful to users of mobile phones. *Journal of Urban Technology, 17*(1), 3–27.

Balk, D., & Yetman, G. (2004). *The global distribution of population: evaluating the gains in resolution refinement.* New York: Center for International Earth Science Information Network (CIESIN), Columbia University.

Ball, P. (2010). Predicting human activity. *Nature, 465*(7299), 692.

Barabási, A.-L. 2010. *Bursts: the hidden patterns behind everything we do, from your e-mail to bloody crusades:* Penguin.

Bhaduri, B., Bright, E., Coleman, P., & Urban, M. L. (2007). LandScan USA: a high-resolution geospatial and temporal modeling approach for population distribution and dynamics. *GeoJournal, 69*(1–2), 103–117.

Birenboim, A., & Shoval, N. (2015). Mobility research in the age of the smartphone. *Annals of the American Association of Geographers, 106*(2), 283–291.

Candia, J., González, M. C., Wang, P., Schoenharl, T., Madey, G., & Barabási, A.-L. (2008). Uncovering individual and collective human dynamics from mobile phone records. *Journal of Physics A: Mathematical and Theoretical, 41*(22), 224015.

Csáji, B. C., Browet, A., Traag, V. A., Delvenne, J.-C., Huens, E., Van Dooren, P., et al. (2013). Exploring the mobility of mobile phone users. *Physica A: Statistical Mechanics and its Applications, 392*(6), 1459–1473.

Cho, E., Myers, S. A, & Leskovec, J. (2011). Friendship and mobility: user movement in location-based social networks. In *Paper read at Proceedings of the 17th ACM SIGKDD International Conference on Knowledge Discovery and Data Mining* (pp. 1082–1090). San Diego, CA: ACM.

Deville, P., Linard, C., Martin, S., Gilbert, M., Stevens, F. R., Gaughan, A. E., et al. (2014). Dynamic population mapping using mobile phone data. *Proceedings of the National Academy of Sciences, 111*(45), 15888–15893.

Dobson, J. E., Bright, E. A., Coleman, P. R., Durfee, R. C., & Worley, B. A. (2000). LandScan: a global population database for estimating populations at risk. *Photogrammetric Engineering and Remote Sensing, 66*(7), 849–857.

Girardin, F., Vaccari, A., Gerber, A., Biderman, A., & Ratti, C. (2009). Towards estimating the presence of visitors from the aggregate mobile phone network activity they generate. In *International Conference on Computers in Urban Planning and Urban Management.*

Gonzalez, M. C., Hidalgo, C. A., & Barabási, A.-L. (2008). Understanding individual human mobility patterns. *Nature, 453*(7196), 779–782.

Harvey, J. T. (2002a). Estimating census district populations from satellite imagery: some approaches and limitations. *International Journal of Remote Sensing, 23*(10), 2071–2095.

Harvey, J. T. (2002b). Population estimation models based on individual TM pixels. *Photogrammetric Engineering and Remote Sensing, 68*(11), 1181–1192.

International Telecommunication Union. (2014). *World Telecommunication Development Conference* (WTDC-2014): Final Report. (ITU, Dubai, United Arab Emirates).

International Telecommunication Union. (2015). *ICT facts and figures—the world* in 2015. (http://www.itu.int/en/ITU-D/Statistics/Documents/facts/ICTFactsFigures2015.pdf last accessed on February 6, 2016).

Isaacman, S., Becker, R., Cáceres, R., Martonosi, M., Rowland, J., Varshavsky, A., Willinger, W. (2012). Human mobility modeling at metropolitan scales. In *Proceedings of the 10th International Conference on Mobile systems, applications, and services* (pp. 239–252). ACM.

Kang, C., Liu, Y., Ma, X., & Wu, L. (2012). Towards estimating urban population distributions from mobile call data. *Journal of Urban Technology, 19*(4), 3–21.

Pei, T., Sobolevsky, S., Ratti, C., Shaw, S.-L., Li, T., & Zhou, C. (2014). A new insight into land use classification based on aggregated mobile phone data. *International Journal of Geographical Information Science, 28*(9), 1988–2007.

Ranjan, G., Zang, H., Zhang, Z.-L., & Bolot, J. (2012). Are call detail records biased for sampling human mobility? *ACM SIGMOBILE Mobile Computing and Communications Review, 16*(3), 33–44.

Ratti, C., Frenchman, D., Pulselli, R. M., & Williams, S. (2006). Mobile landscapes: using location data from cell phones for urban analysis. *Environment and Planning B: Planning and Design, 33*(5), 727–748.

Reades, J., Calabrese, F., & Ratti, C. (2009). Eigenplaces: analysing cities using the space–time structure of the mobile phone network. *Environment and Planning B: Planning and Design, 36* (5), 824–836.

Sevtsuk, A., & Ratti, C. (2010). Does urban mobility have a daily routine? Learning from the aggregate data of mobile networks. *Journal of Urban Technology, 17*(1), 41–60.

Schwanen, T., & Kwan, M.-P. (2008). The internet, mobile phone and space-time constraints. *Geoforum, 39*(3), 1362–1377.

Shanghai Bureau of Statistics. 2014. 2014 年上海市国民经济和社会发展统计公报 [Shanghai Economic and Social Development Statistical Bulletin 2014]. http://www.stats-sh.gov.cn/sjfb/201502/277392.html (last accessed 15 February 2016).

Silm, S., & Ahas, R. (2010). The seasonal variability of population in Estonian municipalities. *Environment and Planning A, 42*(10), 2527–2546.

Soto, V., & Frías-Martínez E. (2011). Automated land use identification using cell-phone records. In *Proceedings of the 3rd ACM International Workshop on MobiArch* (pp. 17–22). ACM.

Stevens, F. R., Gaughan, A. E., Linard, C., & Tatem, A. J. (2015). Disaggregating census data for population mapping using random forests with remotely-sensed and ancillary data. *PLoS ONE, 10*(2), e0107042.

Vieira, M. R., Frias-Martinez, V., Oliver, N & Frias-Martinez, E. (2010). Characterizing dense urban areas from mobile phone-call data: Discovery and social dynamics. In *Paper read at Social Computing (SocialCom), 2010 IEEE Second International Conference on.*

Xu, Y., Shaw, S.-L., Zhao, Z., Yin, L., Fang, Z., & Li, Q. (2015). Understanding aggregate human mobility patterns using passive mobile phone location data: a home-based approach. *Transportation, 42*(4), 625–646.

Xu, Y., Shaw, S.-L., Zhao, Z., Yin, L., Lu, F., Chen, J., et al. (2016). Another tale of two cities: understanding human activity space using actively tracked cellphone location data. *Annals of the American Association of Geographers, 106*(2), 489–502.

Yuan, Y., Raubal, M., & Liu, Y. (2012). Correlating mobile phone usage and travel behavior—a case study of Harbin, China. *Computers, Environment and Urban Systems, 36*(2), 118–130.

Zhao, Z., Shaw, S.-L., Xu, Y., Lu, F., Chen, J., & Yin, L. (2016). Understanding the bias of call detail records in human mobility research. *International Journal of Geographical Information Science, 30*(9), 1738–1762.

Author Biographies

Yang Xu received his Ph.D. from the University of Tennessee, Knoxville. He is currently an Assistant Professor at the Hong Kong Polytechnic University. His research interests include space-time GIS, human mobility mining and modeling, and urban data analytics and visualization.

Shih-Lung Shaw is Alvin and Sally Beaman Professor and Arts and Sciences Excellence Professor of Geography at the University of Tennessee, Knoxville. He also serves as the Interim Associate Provost for international education at the University of Tennessee, Knoxville. He received his B.S. degree from the National Taiwan University and his M.A. and Ph.D. degrees from the Ohio State University. His research interests cover geographic information science (GIScience), transportation geography, time geography, GIS for transportation (GIS-T), and space-time analytics of human dynamics. His recent research has focused on space-time analytics of human activities and interactions in a hybrid physical-virtual world based on various types of individual tracking data such as cell phone data, online social media data, vehicle tracking data, travel-activity survey data, and population migration data. His research has led to the development of a space-time GIS for representation, analysis, and visualization of individual activities and interactions in a hybrid physical-virtual space. Dr. Shaw is a Fellow of the American Association for Advancement of Science (AAAS). He also received the Edward L. Ullman Award for Outstanding Contributions to Transportation Geography from the Association of American Geographers (AAG) and served as the Head of the Department of Geography at the University of Tennessee, Knoxville.

Feng Lu holds a Ph.D. from the Chinese Academy of Sciences. He is a Professor at the Institute of Geographic Sciences and Natural Resources Research, Chinese Academy of Sciences, Beijing, China. His research interests include spatial data management and query, text mining and knowledge graph, and complex network analysis.

Jie Chen received his Ph.D. from the Chinese Academy of Sciences. He is an Assistant Professor at the Institute of Geographic Sciences and Natural Resources Research, Chinese Academy of Sciences, Beijing, China. Her research focuses on human trajectory data analysis and time geography.

Qingquan Li holds a Ph.D. from Wuhan Technical University of Surveying and Mapping. He is a Professor and President of Shenzhen University, China. His research focuses on the integration of GIS, RS and GPS, intelligent transportation, and urban informatics.

Chapter 4
Spatio-Temporal-Network Visualization for Exploring Human Movements and Interactions in Physical and Virtual Spaces

Song Gao, Hanzhou Chen, Wei Luo, Yingjie Hu and Xinyue Ye

4.1 Introduction

With the rapid development of computers, information, and communication technologies, and the increasing availability of mobile phones and social media sources, our living space has been transformed from physical space into a form shared by both physical and virtual spaces (Yu and Shaw 2008; Shaw and Yu 2009). The convergence of geographic information systems (GIS) and social media facilitates the understanding of spatial dynamics of human behavior and societal transformation (Sui and Goodchild 2011). There have been extensive studies on how information (e.g., media contents) diffuses and spread in a social communication network (e.g., Twitter and Facebook). Researchers have developed new theories, methods, and tools for exploring patterns and knowledge hidden in complex human dynamics and social interactions within the context of today's mobile and big data

S. Gao (✉)
Department of Geography, University of Wisconsin, Madison, WI, USA
e-mail: song.gao@wisc.edu

H. Chen
Department of Geography, Pennsylvania State University, State College, PA, USA
e-mail: hzc176@psu.edu

W. Luo
School of Geographical Sciences and Urban Planning, Arizona State University,
Tempe, AZ, USA
e-mail: wluo23@asu.edu

Y. Hu
Department of Geography, University of Tennessee, Knoxville, TN, USA
e-mail: yhu21@utk.edu

X. Ye
Department of Geography, Kent State University, Kent, OH, USA
e-mail: xye5@kent.edu

© Springer International Publishing AG, part of Springer Nature 2018
S.-L. Shaw and D. Sui (eds.), *Human Dynamics Research in Smart and Connected Communities*, Human Dynamics in Smart Cities,
https://doi.org/10.1007/978-3-319-73247-3_4

era (Gao et al. 2013a; Liu et al. 2015; Hu et al. 2015; Shaw et al. 2016, Ye et al. 2016a). Such spatial, temporal and social dynamics can also help discover and understand the rise and fall of emerging topics and events of the society (Shahaf et al. 2012; Peuquet et al. 2015), as well as reveal the digital divide, the physical divide, and the social segregation in developing countries (Amini et al. 2014; Gao et al. 2017). Spatiotemporal visualization techniques are effective in detecting human activity patterns over space and time, while social network analysis is good for exploring graph structures and interactions among individual nodes and groups. Integrating these two types of methods can facilitate the exploration of complex social networks in space and time, and can help discover hidden spatiotemporal and social connections. Luo and MacEachren (2014) proposed a theoretical framework to integrate geographical context, network context, and societal context to understand the geo-social interactions in both physical and virtual spaces. However, few studies so far have made efforts on developing integrated visualization approach and quantitative metrics to measure such interactions in a spatio-temporal-network synthesis framework.

Geo-social network visualization can be classified into two major groups: the first group focuses on spatializing network representation; the second group focuses on exploring geo-social relationships and interactions. The first group tends to integrate network representation into geographical space that has been widely applied to research related to spatial trajectory and interactions, including migration flows (Tobler 1970; Guo 2009), human travel behaviors (Kwan 2004; Gao 2015), transportation flows (Huang et al. 2012), and so on. These studies consider spatial trajectory and interaction from an exclusively spatial perspective without taking their social relationships into account (Shi et al. 2015). Increasingly interconnected societies through technical-social systems (e.g., social media) have called on the necessity of combining social network analysis with spatial analysis (Luo et al. 2011; Luo and MacEachren 2014; Andris 2016; Steiger et al. 2016). Lee and Kwan (2011) integrated 3D space-time paths and ring-based visualization of social networks for analyzing socio-spatial isolation information. Luo et al. (2014) developed a visual analytics tool, the *GeoSocialApp* which supports the exploration of geographical, social network, and multivariable attribute space. Yin and Shaw (2015) proposed a spatio-temporal analysis framework for exploring physical separation and social interactions at the individual level. Luo (2016) developed a visual analytics tool, the *GS-EpiViz*, which allows the exploration of human geo-social interaction patterns to design effective disease control scenarios. These studies proposed several useful visualization framework or analytical approaches for simultaneous discovery of geo-social patterns in physical space or in both physical and virtual spaces. Challenges still remain, however, in quantitatively exploring and measuring relationships between physical movement and social closeness in virtual space.

Researchers have also been actively investigating the spatiotemporal effect in the information diffusion process across a wide range of disciplines (Tsou and Leitner 2013; Ye and Lee 2016). The processes of information diffusion over various network configurations can be analyzed following the approaches in the spatial diffusion studies initiated by the pioneering work of Hägerstrand (1967). Gregory and Urry (1985) argued that Hägerstrand's models cannot deal with spatial diffusion via social networks as well as the associated conflicts and resistance. Morrill et al. (1988) subsequently categorized quantitative models of spatial diffusion processes into stochastic models and deterministic models. Cao et al. (2012) traced the pathways of retweets on a spatial hierarchical layout. Garcia and Wimpy (2016) explored the proliferation of communication technology on information diffusion and violence spread across neighboring states. Spatiotemporal visualization can be applied to intuitively reflect how people share ideas, disseminate information, and communicate with each other, a dynamic process whereby ideas spread through communication networks over time (Tsou et al. 2013). He and Chen (2016) investigated the spatiotemporal text data with visual filters. Novel visualization techniques facilitate the iterative and interactive exploration of the dataset in order to gain deeper understanding of the origination, propagation, and clustering of information (Ye et al. 2016b).

In this research, we propose a conceptual framework for spatiotemporal and social network visualization in a three-dimensional context and present several new quantitative metrics for measuring underlying dynamic interactions among entities. The proposed framework aims to help better understand spatiotemporal patterns of human dynamics and social interactions over both physical and virtual spaces simultaneously, as well as explore how emerging events trigger spatial-temporal-social interactions and information diffusion from a process perspective. As a proof of concept, we demonstrate the proposed framework with a collection of geotagged tweets in the *ArcScene* software. This research provides new insights on integrating multidisciplinary knowledge to explore human dynamics in a broader way.

4.2 An Integrated Spatio-temporal-Network Framework

Here, we introduce a new spatio-temporal-network conceptual framework (STN). As shown in Fig. 4.1, it is an integrated framework which consists of the spatial, temporal, and social context of a subject or an event (labeled as *Nodes*). The spatiotemporal coordinates locate where and when a subject or an event is, while the network linkages (labeled as *Edges*) help know with whom a subject is or where the geo-social impacts of an event are and how the information spread across different people and places via physical movements or the social network connections.

There are three types of edges in the STN. First, the physical edges connect individuals' movements in the physical space (i.e., the space-time path, STP). Second, the social edges represent the individuals' social relationships (e.g., friends and colleagues) or social interactions (e.g., emails, phone calls, retweeting behaviors, and other social media communications) in the virtual space. The subtypes of social edges could vary in different contexts in which at least one or multiple subtypes of social edges could be added in a STN framework instance. Third, the

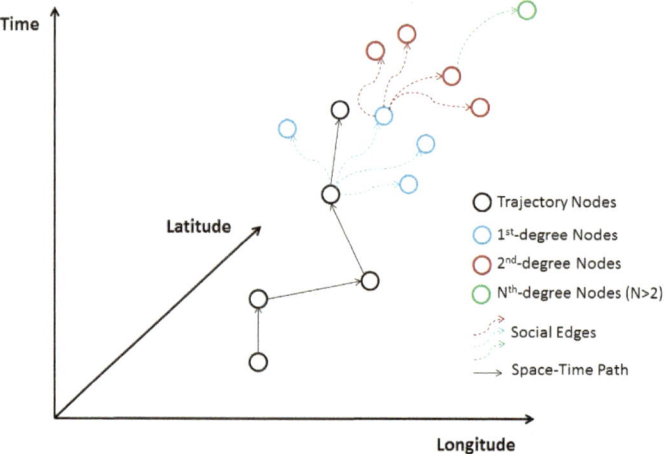

Fig. 4.1 An integrated spatio-temporal-network conceptual framework

physical-social edges connect the individuals' physical locations to their social activities. The third type of edges may or may not be visible in a STN depending on whether the nodes belong to the same type. For example, if the nodes are all individual persons in a STN, we may hide the physical-social edge between an individual and his/her tweet message but only visualize how individuals get connected through retweeting behaviors in the virtual space.

The geographic trajectory nodes which follow a spatiotemporal sequence form the classic STP may form a STN, in which subjects in spatial trajectories are related because of social interactions for a given period triggered by a specific event. The topology or attributes of nodes in such spatial networks can change with time. Information can diffuse through connected nodes in the social network no matter whether nodes are adjacent or not in the physical space. However, the paths (sequential connections) among those spatiotemporal nodes need to obey the time order of the appearance of network edges, which share similar characteristics and inference issues of temporal graphs (Kempe et al. 2000; Holme and Saramäki 2012). The first-degree nodes which are directly connected to the target subject receive the first-hand information, while the second-degree nodes directly connect to and get information from the first-degree nodes, and so forth; the STN has the capability to visualize the physical space-time trajectories and their corresponding Nth-degree virtually connected nodes simultaneously. One potential challenge would be to deal with a large number of individuals and their activities in both physical and virtual spaces. Several visualization techniques can be considered for better visualizing large-scale edges and flows, such as the spatial generalization and aggregation method (Andrienko and Andrienko 2011), the geometry-based edge-clustering framework that can group edges into bundles to reduce the overall edge crossings (Cui et al. 2008), and the hierarchal clustering flow visualization approach (Zhu and Guo 2014).

Human behaviors and information spread by social media has greatly changed the traditional sense of decision-makings. Such a STN can help us visualize and analyze complex spatio-temporal-social interactions. In social networks, the persons who have active interactions in the virtual space may have also had certain connections in the physical space previously. However, how long will the previous physical connections continuously bring virtual interactions? This may depend on the semantics of a person's activities in physical place. For example, attending a conference at a city may have different time decay effect compared with attending college at the same city. The STN may help us visualize and quantify such time decay effect. In addition, this framework can also help with the responses of disaster alerts and evacuation orders. For example, if a wildfire starts, the emergency response staff needs to take action in both physical space and virtual space to announce the evacuation order and the locations of new evacuation shelters, and to reach more people before any further disaster happens and grows.

More specifically, the proposed STN framework may help answer the research questions such as

- How does information diffusion over physical space and virtual space behave differently in normal and crisis situations?
- How do individuals' movements and activities in physical space interact with their social ties in virtual space?
- Which nodes are more influential on specific region and time period?
- What strategies can organizations or governments use to leverage social media and real work channels for decision-makings?
- Can this framework help social scientists trace, monitor, and analyze human dynamics in different domains?
- Which context (spatial, temporal, or social context) is more important to influence information diffusion in different application scenarios?
- How to predict the spread (speed, scale, and range) of social media messages in different spatial-temporal-social networks?

Here, we just list a few potential research questions which could benefit from the proposed STN conceptual framework. Our aim is to stimulate interdisciplinary thinking on research challenges and applications which involve the spatial, temporal, and social aspects.

4.3 Case Study

As a pilot study, we analyzed the recently finished Olympic Games in Rio 2016. This event had drawn attention from people all over the world and generated a large number of topics on social networks, such as Twitter, which can help study the interactions in both physical and virtual space. Often, athletes and reporters in Rio de Janeiro tweeted during the event and interacted with Twitter users from elsewhere in the world by retweeting, commenting, and liking their posts. In particular, we chose to study one of the most influential athletes during this event, Michael Phelps, because his tweets have created a network of connections with people from not only the U.S. but also other countries in the world.

According to Michael Phelps's Twitter account, there were 12 tweets during the time he was in Rio between August 2nd and August 17th. In order to illustrate our STN framework for supporting physical-virtual interactions, we analyzed 3 of his posts: one posted on August 14th was after winning the gold medal of the 4×100 m medley relay, which had the most likes among those 12 tweets; another tweet on August 13th was after he was defeated by a young Singapore athlete Joseph Schooling; and one more post was after he returned back to the U.S. Moreover, we also took Schooling's post on August 15th into account, which was a retweet from Phelps's first tweet on August 13th. The detailed descriptions of these tweets can be found in Table 4.1.

We are interested in analyzing whether there is any geographic discrepancy between the interactions in physical space and those in virtual space over a period of time. Therefore, as picking sample points, we randomly selected 1% of the total tweets. For example, for the tweets on August 14th, we chose 30 posts responding to Phelps's tweets which contain geographic locations and timestamps. From about 2000 retweets of Phelps's tweet on August 13th, we also randomly picked 20 posts related with defined geographic locations and timestamps as well. The collected data were imported into the ArcScene to generate 3D model for visualization.

Table 4.1 Detailed information of four analyzed tweets

User	Description of tweets	Retweet	Time	Location
@MichaelPhelps	What a race!! Congrats to @joschooling !!! We've got an updated pic!! Best of luck bro !! I'll be... https://www.instagram.com/p/BJD6pC2ASbU/	2000	12:48 PM 13 Aug 2016	Rio de Janeiro, Brazil
@MichaelPhelps	Wow! The best way to finish! It has been an honor to represent the USA! A true dream come... https://www.instagram.com/p/BJG362FA58M/	3000	4:22 PM 14 Aug 2016	Rio de Janeiro, Brazil
@joschooling	Thankfully it's a less awkward picture haha! @MichaelPhelps Joseph Schooling added, **Michael Phelps** @MichaelPhelps What a race!! Congrats to @joschooling !!! We've got an updated pic!! Best of luck bro!! I'll be... https://www.instagram.com/p/BJD6pC2ASbU/	171	10:37 AM 15 Aug 2016	Rio de Janeiro, Brazil
@MichaelPhelps	No place like being back home!! Great way to spend my first day in retirement!! boomerrphelps... https://www.instagram.com/p/BJOIk4WgIa0/	1100	12:03 PM 17 Aug 2016	Baltimore, MD

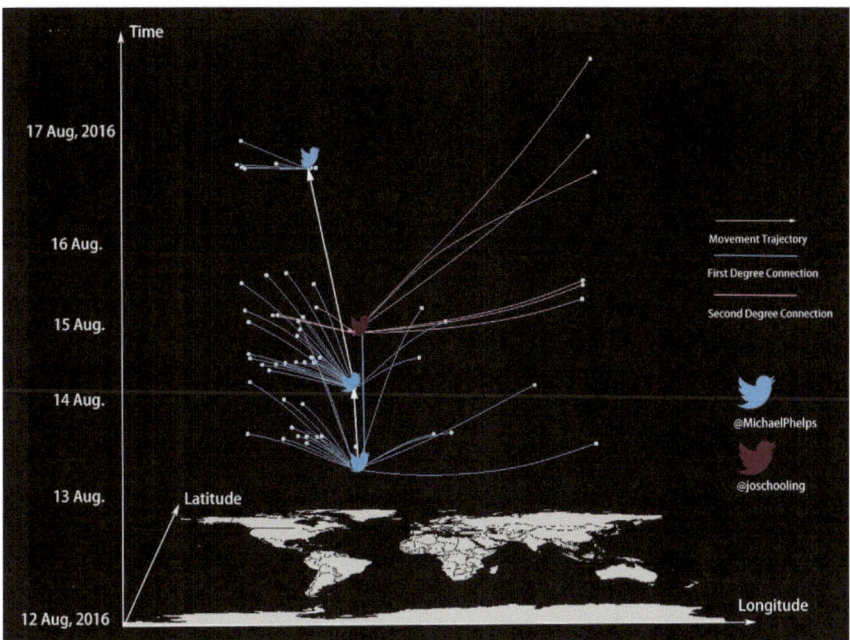

Fig. 4.2   Spatio-temporal-network visualization of geo-social interactions on Twitter

As shown in Fig. 4.2, this 3D space-time-network model created by ArcScene uses x and y coordinates for representing the space, z values for representing time. Points (nodes) represent tweets and polylines for representing either physical movements or social network connections. We acquired x and y coordinates by geolocating tweets to the city-scale. To better illustrate the time in the model, we converted time into minutes. From the visuals, we can see that people do not only interact with people on social network nearby in the physical space, but also interact with those who are far away from them in the physical space. Therefore, when exploring those connections after August 14th, we find out that interactions mostly happened in the United States and few connections with people in Rio or other cities in the world. Moreover, another event relating to Joseph Schooling showed a similar pattern. However, since Schooling is from Singapore, after he interacted with Michael Phelps on Twitter, there were some additional connections to Singapore. Also, as a college student in the University of Texas at Austin, Schooling connected with users near that region as well. In other words, a person's connection with people at a physical place in an earlier time period often brings interactions with the same group of people through virtual space in a later time period.

Moreover, Michael Phelps's tweets have shown a movement trajectory through a scale of time. He moved from Rio back to the U.S., which is indicated by the white arrowed line on the map. As an influential user in virtual space based on the order of time, Phelps's tweets showed a sequence of path with first degree connections that consequently lead to second degree nodes. On top of that, there were more connections generated from users outside of the U.S. during the Olympic, comparing with those after he went back to the U.S.

As for the temporal distribution, we can see that there were more immediate interactions with users living in the U.S. Those lines seemed to create a flatter surface. One possible reason was that there were more users in the U.S. who were more active and paid more attention to Phelps's events. However, as for the connections between users in Asia with Schooling's tweets, they went through a longer time to respond to his post because of time-zone differences, while links to the U.S. showed a similar pattern as that of Michael Phelps's interactions. In short, this case study demonstrated that the STN framework could help understand how events trigger spatio-temporal-social interactions.

4.4 Vision for Quantitative Analytical Metrics

The value of this new conceptual STN framework is in not only considering the art of visualization for designing but also supporting potential development of quantitative analytical indicators for measuring complex interactions among subjects in physical and virtual spaces. Here, we present some of those possible metrics with the STN framework.

- **Spatio-Temporal-Network Impact Extent (STNIE)**

In order to know the spatial impact extent $\{(X_{min}, Y_{min}), (X_{max}, Y_{max})\}$ of a message or an event within a given time period (T_{min}, T_{max}) in a social network, an indicator of the spatio-temporal-network impact extent (STNIE) is defined as:

$$STNIE = \{(X_{min}, Y_{min})_{ti}, (X_{max}, Y_{max})_{tj} | T_{min} \leq ti, tj \leq T_{max})\} \qquad (4.1)$$

The STNIE will be a space-time box in the integrated STN visualization framework. It can help quantify and compare the social impact of an individual node in space and time.

- **Spatio-Temporal-Network Impact Center (STNIC)**

In order to identify the central impact location at a given time period for a given message or an event in the social network, we present an indicator so-called spatio-temporal-network impact center (STNIC). It can have different implementations, such as the weighted mean or the weighted median. The coordinate for weighted mean STNIC is defined as:

$$\bar{X} = \frac{\sum_{i=1} W_i * X_i}{\sum_{i=1} W_i}, \quad \bar{Y} = \frac{\sum_{i=1} W_i * Y_i}{\sum_{i=1} W_i}, \quad \bar{Z} = \frac{\sum_{i=1} W_i * T_i}{\sum_{i=1} W_i} \qquad (4.2)$$

where W_i is the weight for each impacted node (X_i, Y_i, T_i) in the space-time coordinate system. The value of W_i could be the importance of a subject if a node represents a person or could be the number of impacted entities attached to a location if a node in the network represents a place. The weighted mean STNIC might be an arbitrary location which does not exist in the original nodes. Alternatively, by applying the weighted median, a node in the network will be selected as the STNIC. It can be further used for tracking dynamic changes in the spatial distribution of information spread in the virtual space by analogy to human migration center in the physical space (Plane and Rogerson 2015). In our case study, the STNIC for Phelps's tweets locate in the U.S. while that for Schooling is in Asia.

- **Spatio-temporal-Network Distance (STND) and Efficiency (STNE)**

In classic social network analysis, a "path" is a finite sequence of graph edges which connect a sequence of nodes and the network "distance" between two nodes is defined as "the shortest path length" (Wasserman and Faust 1994). Studies on average path length of a network reveal the efficiency of mass transport, information dissemination, or transitive communication on a network. A famous theory of "six-degree separation" describes a chain of a-friend-of-a-friend relations in real world and any two people can be connected within a maximum of six steps (Guare 1990; Watts and Strogatz 1998). A shorter degree of separation has been found in Twitter in which the average path length is 4.12 (Kwak et al. 2010).

In the STN framework, there exist three types of "distance": (1) time delay ($DIST_{time}$); (2) physical movement distance ($DIST_{space}$); (3) path length based on connections on a virtual network ($DIST_{network}$). By analogy to the "graph-based shortest path length", we introduce a spatio-temporal-network distance (STND) which is defined as the "shortest spatio-temporal-network path length" between two nodes with space-time constraints on a STN. Mathematically, it could be expressed as a generic equation below. Furthermore, we can calculate the average spatio-temporal network efficiency (STNE) derived from the STND of all node pairs as a global measure for quantifying the overall efficiency of information spreading on a time-varying network. The formulation for STND and STNE metrics is as follows.

$$STND = \alpha * DIST_{time} + \beta * DIST_{space} + \gamma * DIST_{network} \tag{4.3}$$

$$STNE = \frac{1}{n(n-1)} * \sum_{i,j}^{n} \frac{1}{STND(i,j)} \tag{4.4}$$

where α, β, γ are scale factors for unifying different scales of distance measurements; in application scenarios, one may adjust three scale factors to represent complicated interactions among three different distances. And n represents the total number of nodes in a STN and $STND(i, j)$ is the spatio-temporal-network distance between two nodes i and j.

- **Spatio-temporal-Network Centrality Measures (STNC)**

Traditionally, centrality measures have been employed in determining the relative importance of a central node within a network (Freeman 1977). These measures originate from structural sociology and have been widely applied in studying complex networks, urban structures, and traffic flow (Borgatti 2005; Crucitti et al. 2006; Gao et al. 2013b). Betweenness is a popular type of centrality measures (Freeman et al 1991; Newman 2005) to explore how node importance results from an interaction between the position and the characteristics of the network flow process. However, the classic betweenness measures don't consider the dynamic changes of connections within time-varying networks. In the STN framework, we propose a spatio-temporal-network betweenness centrality (STNBC) measure which can be defined as:

$$STNBC_{i,t} = \sum_{j=1, k=1, i \neq j \neq k}^{n} \frac{N_{j,k}(i,t)}{N_{j,k}(t)} \tag{4.5}$$

where $STNBC(i, t)$ represents the spatio-temporal-network betweenness centrality for a node i at a temporal snapshot t $(T_{min} \leq t \leq T_{max})$. $N_{j,k}(i, t)$ denotes the number of shortest spatiotemporal paths between nodes j and k through node i for a time period $[T_{min}, T_{max}]$, and $N_{j,k}(t)$ is the total number of shortest paths between nodes j and k for the same time period $[T_{min}, T_{max}]$. As mentioned above, those shortest paths which rely on the STND measure may be different because of human movements in physical space or changes of network connections in virtual space over time. It implies that the influence which a node has over the spread of information through the spatiotemporal network may be varying.

4.5 Conclusions

There is a growing recognition of the importance of spatial and temporal dynamic relationships in explaining processes relevant to human behaviors, public health, and social activities (Tsou and Leitner 2013; Ye and He 2016; Ye and Lee 2016). The proposed STN framework can facilitate the understanding of patterns, relationships, and changes in human movements, activities, and communication in the physical-virtual space. This research aims to enable the convergence of new developments in GIS, the art of visualization, social network analytics, and social behavioral research, and facilitate the transformation of social and behavioral science research to computational modeling and analytical applications. It warrants notice that network structures, spatiotemporal and thematic properties, are essential for information spreading. One key contribution of this paper is a novel conceptual framework for integrating space-time visualization and social network analysis for the interaction between physical space and virtual space. To be better prepared for both natural and human-made crises, this framework may be used to facilitate quick spreading of official alerts and warning notifications via multiple virtual-space platforms and real-world channels. We hope that the vision introduced in this article could stimulate interdisciplinary thinking on research challenges and applications which involve the spatial, temporal, and social aspects simultaneously. Those identified research questions and challenges may draw researchers' attention in future work.

References

Adrienko, N., & Adrienko, G. (2011). Spatial generalization and aggregation of massive movement data. *IEEE Transactions on Visualization and Computer Graphics, 17*(2), 205–219.

Amini, A., Kung, K., Kang, C., Sobolevsky, S., & Ratti, C. (2014). The impact of social segregation on human mobility in developing and industrialized regions. *EPJ Data Science, 3*(1), 6.

Andris, C. (2016). Integrating social network data into GISystem. *International Journal of Geographical Information Science, 30*(10), 2009–2031.

Borgatti, S. P. (2005). Centrality and network flow. *Social Networks, 27*(1), 55–71.

Cao, N., Lin, Y. R., Sun, X., Lazer, D., Liu, S., & Qu, H. (2012). Whisper: Tracing the spatiotemporal process of information diffusion in real time. *IEEE Transactions on Visualization and Computer Graphics, 18*(12), 2649–2658.

Crucitti, P., Latora, V., & Porta, S. (2006). Centrality measures in spatial networks of urban streets. *Physical Review E, 73*(3), 036125.

Cui, W., Zhou, H., Qu, H., Wong, P. C., & Li, X. (2008). Geometry-based edge clustering for graph visualization. *IEEE Transactions on Visualization and Computer Graphics, 14*(6), 1277–1284.

Freeman, L. C. (1977). A set of measures of centrality based on betweenness. *Sociometry, 40*, 35–41.

Freeman, L. C., Borgatti, S. P., & White, D. R. (1991). Centrality in valued graphs: A measure of betweenness based on network flow. *Social Networks, 13*(2), 141–154.

Gao, S. (2015). Spatio-temporal analytics for exploring human mobility patterns and urban dynamics in the mobile age. *Spatial Cognition & Computation, 15*(2), 86–114.

Gao, S., Liu, Y., Wang, Y., & Ma, X. (2013a). Discovering spatial interaction communities from mobile phone data. *Transactions in GIS, 17*(3), 463–481.

Gao, S., Wang, Y., Gao, Y., & Liu, Y. (2013b). Understanding urban traffic-flow characteristics: a rethinking of betweenness centrality. *Environment and Planning B: Planning and Design, 40*(1), 135–153.

Gao, S., Yan, B., Gong, L., Regalia, B., Ju, Y., & Hu, Y. (2017). Uncovering the digital divide and the physical divide in senegal using mobile phone data. In *Advances in geocomputation* (pp. 143–151). Cham: Springer.

Garcia, B. E., & Wimpy, C. (2016). Does information lead to emulation? Spatial dependence in anti-government violence. *Political Science Research and Methods, 4*(01), 27–46.

Gregory, D., & Urry, J. (1985). Suspended animation: The stasis of diffusion theory. In D. Gregory & J. Urry (Eds.), *Social relations and spatial structures* (pp. 296–336). New York: St. Martin's Press.

Guare, J. (1990). *Six degrees of separation: A play.* New York: Vintage Books.

Guo, D. (2009). Flow mapping and multivariate visualization of large spatial interaction data. *IEEE Transactions on Visualization and Computer Graphics, 15*, 1041–1048.

Hägerstrand, T. (1967). Aspects of the spatial structure of social communication and the diffusion of information. *Papers in Regional Science, 16*(1), 27–42.

He, J., & Chen, C. (2016, September). Spatiotemporal Analytics of Topic Trajectory. In Proceedings of the 9th International Symposium on Visual Information Communication and Interaction (pp. 112–116). ACM.

Holme, P., & Saramäki, J. (2012). Temporal networks. *Physics Reports, 519*(3), 97–125.

Huang, Z., Das, A., Qiu, Y., & Tatem, A. J. (2012). Web-based GIS: The vector-borne disease airline importation risk (VBD-AIR) tool. *International Journal of Health Geographics, 11,* 1.

Hu, Y., Gao, S., Janowicz, K., Yu, B., Li, W., & Prasad, S. (2015). Extracting and understanding urban areas of interest using geotagged photos. *Computers, Environment and Urban Systems, 54,* 240–254.

Kempe, D., Kleinberg, J., & Kumar, A. (2000, May). Connectivity and inference problems for temporal networks. In *Proceedings of the thirty-second annual ACM symposium on Theory of computing* (pp. 504–513). USA: ACM.

Kwan, M. P. (2004). GIS methods in time-geographic research: Geocomputation and geovisualization of human activity patterns. *Geografiska Annaler: Series B, Human Geography, 86,* 267–280.

Kwak, H., Lee, C., Park, H., & Moon, S. (2010). What is Twitter, a social network or a news media?. In *Proceedings of the 19th international conference on World wide web,* (pp. 591–600). ACM, April.

Lee, J. Y. & M. P. Kwan (2011). Visualisation of socio-spatial isolation based on human activity patterns and social networks in space-time. *Tijdschrift voor economische en sociale geografie, 102,* 468–485.

Liu, Y., Liu, X., Gao, S., Gong, L., Kang, C., Zhi, Y., et al. (2015). Social sensing: A new approach to understanding our socioeconomic environments. *Annals of the Association of American Geographers, 105*(3), 512–530.

Luo, W. (2016). Visual analytics of geo-social interaction patterns for epidemic control. *International Journal of Health Geographics, 15,* 28.

Luo, W., & MacEachren, A. M. (2014). Geo-social visual analytics. *Journal of Spatial Information Science, 2014*(8), 27–66.

Luo, W., MacEachren, A. M., Yin, P., & Hardisty, F. (2011, November). Spatial-social network visualization for exploratory data analysis. In *Proceedings of the 3rd ACM SIGSPATIAL International Workshop on Location-Based Social Networks* (pp. 65–68). USA: ACM.

Luo, W., Yin, P., Di, Q., Hardisty, F., & MacEachren, A. M. (2014). A geovisual analytic approach to understanding geo-social relationships in the international trade network. *PLoS ONE, 9,* e88666.

Morrill, R., Gaile, G. L., & Thrall, G. I. (1988). Spatial diffusion. SAGE Scientific Geography Series 10. Newbury Park, CA: SAGE Publications, Inc.

Newman, M. E. (2005). A measure of betweenness centrality based on random walks. *Social networks, 27*(1), 39–54.

Peuquet, D. J., Robinson, A. C., Stehle, S., Hardisty, F. A., & Luo, W. (2015). A method for discovery and analysis of temporal patterns in complex event data. *International Journal of Geographical Information Science, 29*(9), 1588–1611.

Plane, D. A., & Rogerson, P. A. (2015). On tracking and disaggregating center points of population. *Annals of the Association of American Geographers, 105*(5), 968–986.

Shahaf, D., Guestrin, C., & Horvitz, E. (2012, April). Trains of thought: Generating information maps. In *Proceedings of the 21st international conference on World Wide Web* (pp. 899–908). USA: ACM.

Shaw, S. L., Tsou, M. H., & Ye, X. (2016). Editorial: Human dynamics in the mobile and big data era. *International Journal of Geographical Information Science, 30*(9), 1687–1693.

Shaw, S. L., & Yu, H. (2009). A GIS-based time-geographic approach of studying individual activities and interactions in a hybrid physical–virtual space. *Journal of Transport Geography, 17*(2), 141–149.

Shi, L., Chi, G., Liu, X., & Liu, Y. (2015). Human mobility patterns in different communities: a mobile phone data-based social network approach. *Annals of GIS, 21*(1), 15–26.

Steiger, E, Westerholt, R & Zipf, A. (2016). Research on social media feeds—A GIScience perspective. In: Capineri, C, Haklay, M, Huang, H, Antoniou, V, Kettunen, J, Ostermann, F and Purves, R. (eds.) *European handbook of crowd sourced geographic information* (pp. 237–254). London: Ubiquity Press. http://dx.doi.org/10.5334/bax.r. License: CC-BY 4.0.

Sui, D., & Goodchild, M. (2011). The convergence of GIS and social media: Challenges for GIScience. *International Journal of Geographical Information Science, 25*(11), 1737–1748.

Tobler, W. (1970). A computer movie simulating urban growth in the Detroit region. *Economic Geography, 46,* 234–240.

Tsou, M. H., & Leitner, M. (2013). Editorial: Visualization of social media: Seeing a mirage or a message? In special content issue: "Mapping cyberspace and social media". *Cartography and Geographic Information Science., 40*(2), 55–60.

Tsou, M. H., Kim, I. H., Wandersee, S., Lusher, D., An, L., Spitzberg, B., Gupta, D., Gawron, J. M., Smith, J., Yang, J. A., & Han, S. (2013). Mapping Ideas from cyberspace to real space: Visualizing the spatial context of keywords from web page search results. *International Journal of Digital Earth, 7*(4), 316–335.

Wasserman, S., & Faust, K. (1994). *Social network analysis: Methods and applications* (Vol. 8). Cambridge: Cambridge university press.

Watts, D. J., & Strogatz, S. H. (1998). Collective dynamics of 'small-world' networks. *nature, 393* (6684), 440–442.

Ye, X., & He, C. (2016). The new data landscape for regional and urban analysis. *GeoJournal.* https://doi.org/10.1007/s10708-016-9737-8.

Ye, X., Huang, Q., & Li, W. (2016a). Integrating big social data, computing, and modeling for spatial social science, cartography and geographic information science. *Science, 43*(5), 377–378.

Ye, X., & Lee, J. (2016). Integrating geographic activity space and social network space to promote healthy lifestyles. *ACMSIGSPATIAL Health GIS, Newsletter, 8*(1), 24–33.

Ye, X., Li, S., Yang, X., & Qin, C. (2016b). Use of social media for detection and analysis of infectious disease in China. *ISPRS International Journal of Geo-Information.* https://doi.org/10.3390/ijgi5090156.

Yin, L., & Shaw, S. L. (2015). Exploring space–time paths in physical and social closeness spaces: a space–time GIS approach. *International Journal of Geographical Information Science, 29*(5), 742–761.

Yu, H., & Shaw, S. L. (2008). Exploring potential human activities in physical and virtual spaces: A spatio-temporal GIS approach. *International Journal of Geographical Information Science, 22*(4), 409–430.

Zhu, X., & Guo, D. (2014). Mapping large spatial flow data with hierarchical clustering. *Transactions in GIS, 18*(3), 421–435.

Author Biographies

Song Gao (Ph.D., University of California, Santa Barbara), Assistant Professor at the Department of Geography, University of Wisconsin, Madison. His main research interests include Place-Based GIS, Spatiotemporal Data Analytics and Geospatial Semantics. He utilizes multi-data sources with spatio-temporal-network analysis methods to study human mobility patterns and urban spatial structure.

Hanzhou Chen (MA/Ph.D. Student, Pennsylvania State University), Research Assistant in Friendly Cities Lab at Penn State. His interests are in human mobility, social networks, and social flows. He is especially interested in studying and visualizing human mobility and interactions within urban and within a spatial and temporal framework.

Wei Luo (Ph.D., Pennsylvania State University), Postdoc at the Arizona State University. His research interests include geovisual analytics and large-scale location-based social networks focusing on human dynamics and disease transmission and control (e.g., influenza, HIV), climate change impacts, international trade, social media, and criminal activity.

Yingjie Hu (Ph.D., University of California, Santa Barbara), Assistant Professor, University of Tennessee, Knoxville. The main research area of Yingjie Hu is Geographic Information Science and more specifically Geospatial Semantics. His research aims to bring a semantic perspective which enables intelligent systems to automatically understand geographic information and helps explain spatiotemporal patterns of geographic phenomena. His interest in human dynamics focuses on the relations between the textual words expressed by people and their behavior in space and time.

Xinyue Ye (Ph.D., University of California at Santa Barbara and San Diego State University), is an associate professor in the Department of Geography at Kent State University where he directs computational social science lab. His major expertise is on modelling the geographical perspective of socioeconomic inequality and human dynamics. He develops and implements new methods on spatiotemporal-social network analysis/modelling/simulation for different application domains such as economic development, disaster response, land use, public health, and urban crime. He has received about 4.5 million dollars in grants as PI or Co-PI from National Science Foundation, Department of Commerce, and Department of Energy.

Chapter 5
Modeling Mobility and Dynamics of Scheduled Space-Time Activities—An RDF Approach

Junchuan Fan and Kathleen Stewart

5.1 Introduction

The rapid advancements of information and communication technologies (ICT) have dramatically changed the way people schedule and conduct daily space-time activities. The ubiquity of location-enabled devices has enabled people to publish and receive information about their surrounding environment, in which they are also acting as sensors themselves (Goodchild 2007). The increasing availability of data about individual activities has contributed to our understanding of human mobility and urban dynamics based on activity data. Although activity-based analysis can provide a disaggregated perspective that may capture more nuanced impact factors (e.g., scheduling constraints, links among activities) that better determine human mobility than aggregated trip-based approaches, traditional geographic information systems (GIS) data models and analytic frameworks do not work well for activity-based analyses (Miller 2014). Conventional geographic data models and information systems fall short when dealing with the unique characteristics of individual activity data, namely, the large volume, the high variety of data sources, and the velocity of the generated activity data (Miller and Goodchild 2014). The relational data model, the most common logical data model used by many GIS, is not flexible enough, especially when additional dimensions (e.g., social network connections between people) are considered. On the other hand, the idea of how to make our daily environment smart with the help of ICT is an important research topic, e.g., building smart cities using sensor networks (Galache et al. 2013), deploying cameras for intelligent transportation networks

J. Fan (✉) · K. Stewart
Department of Geographical Sciences, University of Maryland, College Park, MD, USA
e-mail: jcfan@umd.edu

K. Stewart
e-mail: stewartk@umd.edu

© Springer International Publishing AG, part of Springer Nature 2018
S.-L. Shaw and D. Sui (eds.), *Human Dynamics Research in Smart and Connected Communities*, Human Dynamics in Smart Cities,
https://doi.org/10.1007/978-3-319-73247-3_5

(Calderoni et al. 2014) and modeling contextual information of indoor environments to facilitate navigation (Afyouni et al. 2013) and smart campuses (Sengupta et al. 2010; Wang and Ng 2012). The latter could involve for example, the integration of mobile ICT into the implementation of smart campuses; better understanding of diurnal and seasonal demographics of campus buildings and spaces using geospatial-enabled ICT; and additional studying, teaching, and research opportunities fostered by smart campus information system and infrastructure. The field of geodesign has similar interests through efforts to disseminate spatial thinking into the planning and designing profession, facilitating the creation of context-sensitive smart environments (Abukhater and Walker 2010).

In this chapter, we build on these perspectives to propose a semantic data modeling framework based on semantic web technologies for representing, querying, reasoning, and visualizing human movements that are linked to scheduled space-time activities, specifically the movements of students as they follow weekly course schedules and move about a campus. The foundation of semantic web technologies is the Resource Description Framework (RDF[1]), a W3C standard data modeling framework for semantically describing resources on the web. RDF represents information about resources in graph form. Information about resources are represented by triples, <*subject, predicate, object*>, linked together through a chain of predicates. A set of RDF triples forms a *triple store*, or a *knowledgebase*. Web ontology language (OWL[2]), a widely used knowledge representation language based on description logic, extends the semantic modeling expressiveness of RDF by incorporating more nuanced semantic constructs (e.g., *functional* and *inverse relationships*). Under this framework, space-time activities specifically, scheduled course activities on campus, are represented as ontological classes that have both the characteristics of spatial objects and temporal entities, enabling the semantic reasoner to infer the spatiotemporal relationships among different space-time activities. Domain entities (for example, *scheduled course, student* and *campus facility*) are modeled as a set of ontological classes formally expressed using OWL. The semantic relationships among domain entities (for example, *hasParticipant, hasTrajectory*) are captured through semantic predicates.

The main contributions of this chapter are: (1) a semantic data model for space-time activities based on semantic web technologies that integrates spatial, temporal and semantic aspects of space-time activities and movements; (2) a ontological framework that links space-time activities with their participants through a hierarchy of thematic roles, and thus facilitating the reasoning of movement dynamics at different granularities; (3) a unified approach for reasoning about and retrieving movement dynamics from both aggregated and individual perspectives.

Course scheduling information for The University of Iowa, in Iowa City, IA during fall 2014 is adopted and transformed into a spatiotemporal RDF

---

[1]https://www.w3.org/RDF/.

[2]https://www.w3.org/OWL/.

knowledgebase. Using this prototype system, we discuss an approach for querying, analyzing and visualizing students' mobility patterns as they follow their daily/ weekly course schedules, and the aggregated spatial and temporal dynamics for a campus (e.g., space usage, busy hours). The rest of this chapter is organized as follows: Sect. 5.2 discusses related work on human mobility research, and geospatial semantic research; Sect. 5.3 presents a semantic modeling framework for modeling space-time activities that integrates the spatial, temporal and semantic dimensions; Sect. 5.4 discuss a system framework for developing a prototype system based on a university course scheduling information; Sect. 5.5 presents a use case using course schedules from the University of Iowa, querying and visualizing the activity spaces of students as well as the inflow of students to buildings; the final section presents conclusions and a discussion about future work.

5.2 Related Work

In this chapter, the domain application is the movement of students participating in scheduled activities (e.g., courses and seminars) on a campus. Related work on space-time mobility, spatiotemporal scheduling, and geospatial semantic research, will be discussed.

5.2.1 Geospatial Technologies and Human Mobility

A large body of work on human movement in response to daily activities, including theories, methods and tools, have been developed by the time geography community (Miller 1991; Kwan et al. 2003; Miller 2005; Yu 2006; Shaw and Yu 2009; Andrienko et al. 2013a). The theoretic framework for measuring human space-time accessibility proposed by Hägerstrand (1970) has been extended in many aspects as GIS technologies have advanced. In contrast to early assumptions that individuals move in space without constraint, street networks are frequently incorporated into the measurement of space-time accessibility (Miller 1999). The human movement space can also be expanded from physical space considerations only to include virtual spaces to accommodate changes brought about by rapid progress in information and communication technologies. Spatiotemporal GIS, for example, has been designed in order to explore the individual activities and interactions in a hybrid physical-virtual space (Yu 2006; Shaw and Yu 2009; Yin and Shaw 2015). Equipped with location-aware handheld devices, individuals are now acting as sensors (Goodchild 2007) being aware of the physical as well as social context in which they are situated. With the rapid progress of cloud and mobile computing technologies, individuals now have the capability to understand what's going on around them in real time (Crooks et al. 2013; Majid et al. 2013) and alter their daily scheduled activities accordingly (Stewart et al. 2013). Research on mining complex human mobility patterns from massive trajectory data further enhance our understanding of human movement and the interactions of people with their surrounding environments (Andrienko et al. 2013a, b; Giannotti et al. 2011). A formal model about individual activities scheduling is presented in (Stewart et al. 2013). This model goes beyond the conventional organization of scheduled activities that provides only a temporal view of a schedule, and represents both the spatial and temporal aspects of an individual's scheduled activities. A set of schedule operations including *reschedule, postpone, change location,* and *delete* are modeled in a task ontology, affording a scheduler system that can perform semantic reasoning with reference to an underlying ontology.

In this chapter, semantic web technologies are used for analyzing space-time movements associated with scheduled activities on a campus. This extends the time geographic theoretic framework to include domain ontologies of scheduled activities. Semantics associated with space-time movement of individuals on a campus can be exploited to improve campus planning as well as the interactions of individuals as they go about their daily pursuits and undertake their planned activities.

5.2.2   Semantic Data Models and Geospatial Research

The availability of geospatial data and services has increased due to the rapid development of information and communication technologies, especially pervasive handheld devices, such as cellphones equipped with positioning capabilities. As a response to the need for intelligent sharing and processing of geospatial information and services, geospatial semantics have emerged as an important research area. Geospatial semantic research is a sub field of semantic web research that deals with location-based information and semantics, namely the meaning of geographic concepts and their underlying interrelationships. These semantics capture that different conceptualizations that can hold for geographic phenomenon that are often apparent from the natural language that people use to describe geographic features; for example, 'bank' can have different meanings under different contexts (Kuhn 2012). In addition to supporting different meanings, semantic interoperability may be desirable, for example, geospatial information may be needed at multiple spatiotemporal granularities. Kuhn (2005) discussed the need to formally define semantics of geospatial concepts and construct semantic reference systems in order to support semantic interoperability. Inspired by spatial reference systems, a framework was presented for referencing, grounding and mapping geospatial information in the form of a semantic reference system. Using geospatial ontologies to capture the particular semantics of geospatial features and relations can help users retrieve geospatial information and services more effectively, especially as the traditional web of documents is turning into a web of data (Egenhofer 2002; Janowicz et al. 2012).

Besides semantic interoperability, the ability to perform semantic reasoning using semantic data models is another important incentive for research in this area. Thematic relationships among geospatial features, aside from spatiotemporal relationships, are increasingly exploited in different kinds of analytic applications, for example, in a national security application where detecting conflicts of interest were studied (Perry et al. 2007). In this work, a military ontology is combined with an upper-level ontology to model interactions relating to military combat. Thematic entities and relationships are modeled as first class objects, linked to spatial entities through *located_at* and *occurred_at* relationships from the upper-level ontology. For example, a soldier is associated with a training facility using a set of relationships (*Soldier—member_of—Military_Unit—trains_at—Base—located_at—Spatial_Entities*) (Perry 2008). Enriching spatiotemporal data with semantic information abstracts the data from geographic space to semantic space (Andrienko et al. 2013a), providing the means to investigate and discover more general patterns of geographic dynamic phenomena. This is especially important in human movement data analysis where spatiotemporal behavior patterns are of particular interest to researchers.

5.3 A Semantic Data Model for Space-Time Activity

5.3.1 Integrate Spatial and Temporal Dimension of Space-Time Activity

Time geography research investigates human mobility in a 3D space-time conceptual model (Neutens et al. 2008; Shaw and Yu 2009; Chen et al. 2011; Crease and Reichenbacher 2013), where the *x* and *y* dimensions represent the spatial aspect and *z* dimension represents time. Individuals move about in both space and time to undertake daily activities, referred to as *space-time activities* in this research. These are the main driving force for human movement, and provide a unique perspective for studying human mobility. From a formal model perspective, a space-time activity can be represented as an entity that has attributes of both a spatial object and a temporal entity (Fig. 5.1), and in the domain ontology developed for this research, space-time activity (*ST_Activity*) is modeled as a class that inherits properties from both the *SpatialObject* class defined in GeoSPARQL and the *TemporalEntity* class defined in the W3C time ontology. GeoSPARQL is an OGC

Fig. 5.1 Space-time activity modeled as both *SpatialObject* and *TemporalEntity*

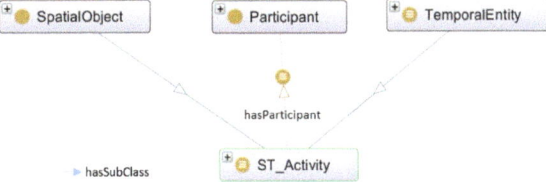

standard designed to support representation and queries about geospatial objects, including qualitative reasoning about topological (i.e., spatial) relationships between spatial objects (e.g., nine-intersection topological relations). The W3C time ontology on the other hand, defines properties of temporal entities and relations between them [i.e., Allen's temporal interval relation algebra (Allen 1984)].

Formally, we can define a space-time activity entity using OWL and formalized in *turtle format* as:

```
@prefix time:  <http://www.w3.org/2006/time#> .
@prefix geo:   < http://www.opengis.net/ont/geosparql#> .
@prefix rdfs:  <http://www.w3.org/TR/rdf-schema/#> .
@prefix :      <http://st-activity.org/#> .
:st_activity rdfs:subClassOf geo:SpatialObject .
:st_activity rdfs:subClassOf time:TemporalEntity.
```

A temporal entity can be modeled as either a *time:Instant* or *time:Interval*, and in this way, space-time activities can be handled at different temporal granularities, which is important for representing different types of activities (Crease and Reichenbacher 2013). By integrating attributes of both *geo:SpatialObject* and *time: TemporalEntity*, a semantic web reasoning engine can be applied to reason about the spatiotemporal relationships between space-time activities and movements.

5.3.2   Thematic Role of Participants of Space-Time Activities

The minimally structured and linked nature of RDF data model of space-time activities provides researchers with great flexibility in terms of analyzing human movement dynamics from different perspectives (e.g., individual movement path, spatial pattern, temporal fluctuation) (Andrienko et al. 2011). In this section, we discuss the thematic roles of *participants* in space-time activities based on research from knowledge representation community. The link between space-time activities and their participants through different thematic roles is important for retrieving human movement dynamics at different granularities based on activity data.

In knowledge representation research, thematic roles are used to represent the links between an *occurrent* and its *participants*. From an ontological perspective, space-time activities are *occurrents* (Grenon and Smith 2004) that are ephemeral, i.e., they happen and then no longer exist. Space-time activities involve different types of participants (e.g., individual students, faculty, and classroom) and every participant is an entity that plays some role during an activity. The relation of *participation* holds between a substance and a process. Smith and Grenon (2004) discussed different modes of *participation* between a SNAP entity (a continuant) and a SPAN entity (an occurrent) (for example, *initiation, perpetuation, termination, facilitation, hindrance, mediation*). In knowledge representation and artificial

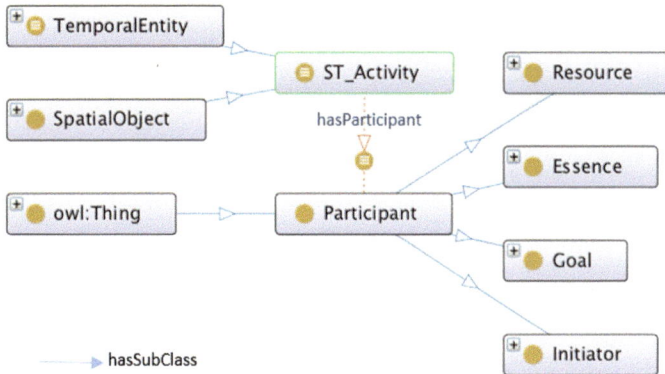

Fig. 5.2 Participants of space-time activity

intelligence community, there are also abundant researches on thematic roles of entities during a process. Sowa (1999) classified thematic roles for participants in a process into four categories in accordance with Aristotle's four causes, namely, *initiator, resource, goal, and essence* (Fig. 5.2). These thematic roles capture fundamental relationship semantics between *occurrents* and their *participants*, complementing Smith and Grenon's classification, and can be used as building blocks for definition of higher level semantics like Smith and Grenon's. We build on Sowa's classification of thematic roles for the different types of participants in space-time activities.

In the semantic model of space-time activities, *initiator* is the agent that starts an activity with voluntary intention (for example, faculty who offer courses), and *resource* must be present throughout the activity, but does not actively control the event (e.g., the venue being used for a course activity). *Goal* is a participant in an activity that represents the purpose of the activity and controls the activity from the reverse direction (e.g., students who take courses). *Essence* refers to an essential participant of an activity that represents the theme or byproduct of an activity (e.g., the discussion topic of a course activity). Different types of participants will participate in different kinds of space-time activities. It is not necessary that all four types of participants are present in each space-time activity. Out of the four main types of participants for a space-time activity, *Initiator and Resource* are the active determinants of the activity, whereas *Goal* and *Essence* are the product or experience of the activity. In other words, *Initiator* and *Resource* participants model the semantic constraints that need to be satisfied in order for a space-time activity to successfully take place, while *Goal* participant represents the end results or recipients of a space-time activity.

For example, the fact that a student named Mary is attending a scheduled course, Foundation of GIS in :JH243(Jessup Hall 243) on campus is formalized in the RDF knowledgebase as:

```
:Foundation_of_GIS_A3 a  :st_activity.
:Foundation_of_GIS_A3 time:hasBeginning :f_begin .
:f_begin a time:Instant;
:inXSDDateTime "2014-09-19T11:30:00"^^xsd:dateTime .
:Foundation_of_GIS_A3 time:hasEnd :f_end .
:f_end a time:Instant;
:inXSDDateTime "2014-09-19T12:20:00"^^xsd:dateTime .
:Foundation_of_GIS_A3 geo:hasExactGeometry [
        geo:asWKT "Polygon((-91.53639 41.6622, -91. 53663 41.6622,
-91.53663 41.6616, -91.53639 41.6616)"^^geo:wktLiteral ].
:Foundation_of_GIS_A3 :hasRecipient :Mary.
:Foundation_of_GIS_A3 :hasInitiator  :YJ.
:Foundation_of_GIS_A3 :hasPhysicalSpace :JH243.
```

In the RDF triples listed above, both :*Mary* and :*YJ* are instances of *foaf:Person* class from FOAF[3] ontology. The :*Foundation_of_GIS_A3* course is an instance of :*st_activity*, and has several different types of participants that play different thematic roles. If there is no participant, then an :*st_activity* has no impact in space-time and does not need to be included in the knowledgebase. The most granular spatial information for this course (i.e., the boundary of the building in which this course is scheduled) is its geometry, serialized as well-known text (WKT) format. :JH243. This spatial footprint is an instance of *geo:SpatialOBject* class, and it is linked with :*Foundation_of_GIS_A3* via :*hasPhysicalSpace* predicate, which is a sub-property of :*hasResource*. :*YJ* is the instructor for this course and therefore the initiator. : *Mary* is one of the students that attend this course, and thus the *recipient* of the course. Note that each type of participants can be further classified into more specific roles, and they follow a hierarchical structure (e.g., *PhysicalSpace* **rdfs:subClassOf** *Resource*). As a result, the semantic predicates linking space-time activities and their participants are also hierarchical (*hasPhysicalSpace* **rdfs:subPropertyOf** *hasResource*), giving us the flexibility to query and analyze movement dynamics at different granularities.

5.3.3   Individual Trajectory and Space-Time Activity

In our semantic modeling framework, the set of space-time activities that a moving agent undertakes forms a *trajectory*. *Trajectory* is a subclass of *geo:SpatialObject* (Fig. 5.3), and therefore it is associated with a spatial footprint. Space-time activity is part of a trajectory, and participants that belong to *foaf:Person* class are linked with the trajectory class through :*hasTrajectory* property.

---

[3]http://xmlns.com/foaf/spec/.

Fig. 5.3 Space-time activity part of a trajectory

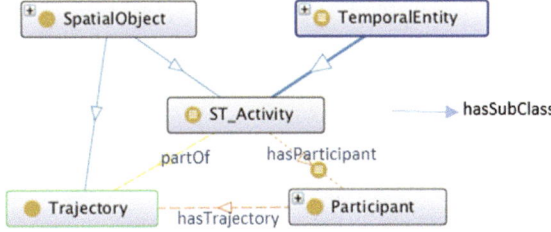

5.3.4 Scheduled Space-Time Activities

The movements of individuals on a campus are driven by locations and times of courses that are being offered. These *scheduled* space-time activities are a subclass of space-time activity that satisfies certain spatial and/or temporal constraints. For this research, we characterize scheduled space-time activities, as being spatially fixed temporally fixed (SXTX). A spatially fixed temporally fixed (SXTX) activity have set start and end times as well as set locations. People who participate in SXTX activities have to satisfy the constraints that they need to travel to the preset location and be present during the activity. Most courses fall into this category, for example, students who are enrolled in the Foundations of GIS course are assumed to be in the classroom during the lecture. Regular Courses on campus are SXTX activities and inherit all the properties of SXTX activity class. In the next section we discuss our prototype system that is used to capture movement dynamics reflected by students undertaking SXTX activities.

5.4 Prototype Implementation and a Use Case for Scheduled Movements

Traditionally, a university campus has at least one information system that manages information about students and the courses they undertake for a degree. Although spatial and temporal information about scheduled activities (e.g., courses, seminars etc.) are available, users of the information system typically only view the information in a text format. In this section, the semantic modeling framework is demonstrated through a prototype system using the University of Iowa campus as the study domain. There are four components for the prototype system (Fig. 5.4).

Building a smart campus will benefit from a spatial enablement of the environment (Roche 2014), and this is the first component of the framework. This process takes as input scheduled course information encoded as plain text and adds a spatial dimension. The location information associated with scheduled activities is geocoded and transformed into geographic coordinates. The second component of the framework involves a semantic enrichment process where a semantic dimension is added to the spatially-enabled scheduled activities, generating an RDF

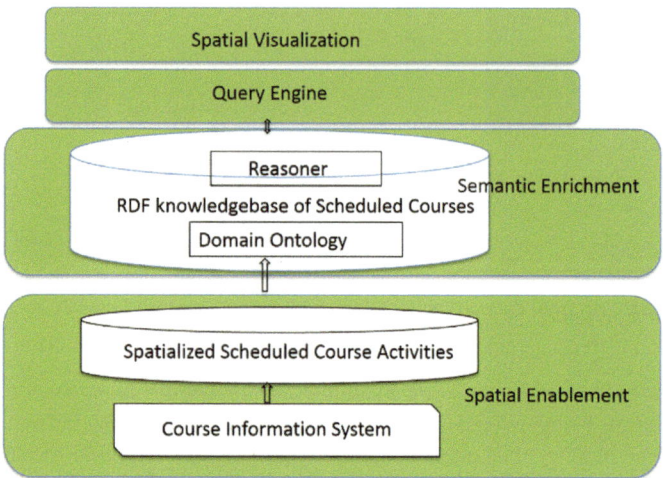

Fig. 5.4 Architecture of prototype system for campus course activities

knowledgebase for scheduled course activities. This component includes a domain ontology and a reasoner that can perform spatial, temporal, and semantic reasoning on scheduled activities. The reasoner is built in the RDF knowledgebase, inferring implicit relationships among domain entities based on the developed ontology. The third component of the framework is a GeoSPARQL query engine that can parse movement-related queries and send them to the RDF knowledgebase. The final component of this framework is a web-based visualization interface for the query results, reflecting the movement dynamics on campus.

The course information for fall semester 2014 listed on The University of Iowa online information system (https://myui.uiowa.edu/my-ui/courses/dashboard.page) is employed as a data source for this study. A Python program has been developed that scrapes all course information off the HTML webpage. The scraped data are organized and transformed into RDF triples using RDF4J library. The temporal reasoning module is realized by integrating SWRLAPI,[4] an open source java package that uses the semantic web rule language. The spatial reasoning module is developed based on open source java packages spatial4j[5] and JTS.[6] There are on average 2000 courses held each weekday during the fall term. An ArcMap shapefile containing campus buildings provides spatial reference for the scheduled campus activities. To test what movements arise relating to different course schedules,

---

[4]https://github.com/protegeproject/swrlapi.

[5]https://github.com/locationtech/spatial4j.

[6]https://github.com/metteo/jts.

the course schedules of 13 students registered in fall 2014 from four different majors, including computer science, geography, environmental science, and geology were collected. For example, student *A* is an Environmental Science major registered for five courses. Another student *B*, is a Geography major registered for six courses including *Creative writing studio workshop, Foundation of GIS, Foundation of GIS A03 (lab section), The Global Economy, Environment Justice, Civilization of Asia: South Asia,* and *Relaxation Techniques.*

5.5 Retrieving Campus Movement Dynamics from RDF Knowledgebase

5.5.1 *Representing and Visualizing* Trajectories *for Individual Students*

In general, students move around campus while they attend the courses for which they are registered. The sequence of activities as they move from one classroom to another forms a *Trajectory* for the student. Since scheduled course activities have both spatial and temporal characteristics, visualization of the movement trajectory associated with attending these scheduled course activities can be generated. In order to generate realistic trajectories representing students' movements, road network information is incorporated into the RDF knowledgebase. The movement trajectories of students *A* and *B* are calculated using a shortest path routing algorithm on the campus road network. In both cases, the generated trajectories respect barriers such as a river. The movement trajectory for Student *A* is based on an aggregate of 10 scheduled class sessions (each course has two or three sessions per week, thus 10 sessions for 5 courses) (Fig. 5.5a) while Student *B*'s schedule is based on an aggregation of 15 class sessions (Fig. 5.5b). The travelling distance of *A* is longer than that of *B*, as *A* is an environmental science student who is also interested in Art. This student regularly travels across campus and take classes in the Studio Arts building that is more than a mile from the main campus. The different movement patterns, when combined with geographic contextual information (e.g., the functions of different parts of the campus, (e.g., the Main Library and Liberal Arts and Science buildings), reveal the academic as well as extracurricular interests of students. Locations of courses are shown with purple symbols. The spatial visualization of movement trajectories provide an understanding about student activity spaces. The areas that students move around in most frequently are identified based on a temporal aggregation of the weekly or monthly trajectories. Using such an approach makes it possible for students to potentially optimize decisions about locations for studying, getting coffee or lunch, and hanging out with friends in relation to the location of courses they are taking.

(a)

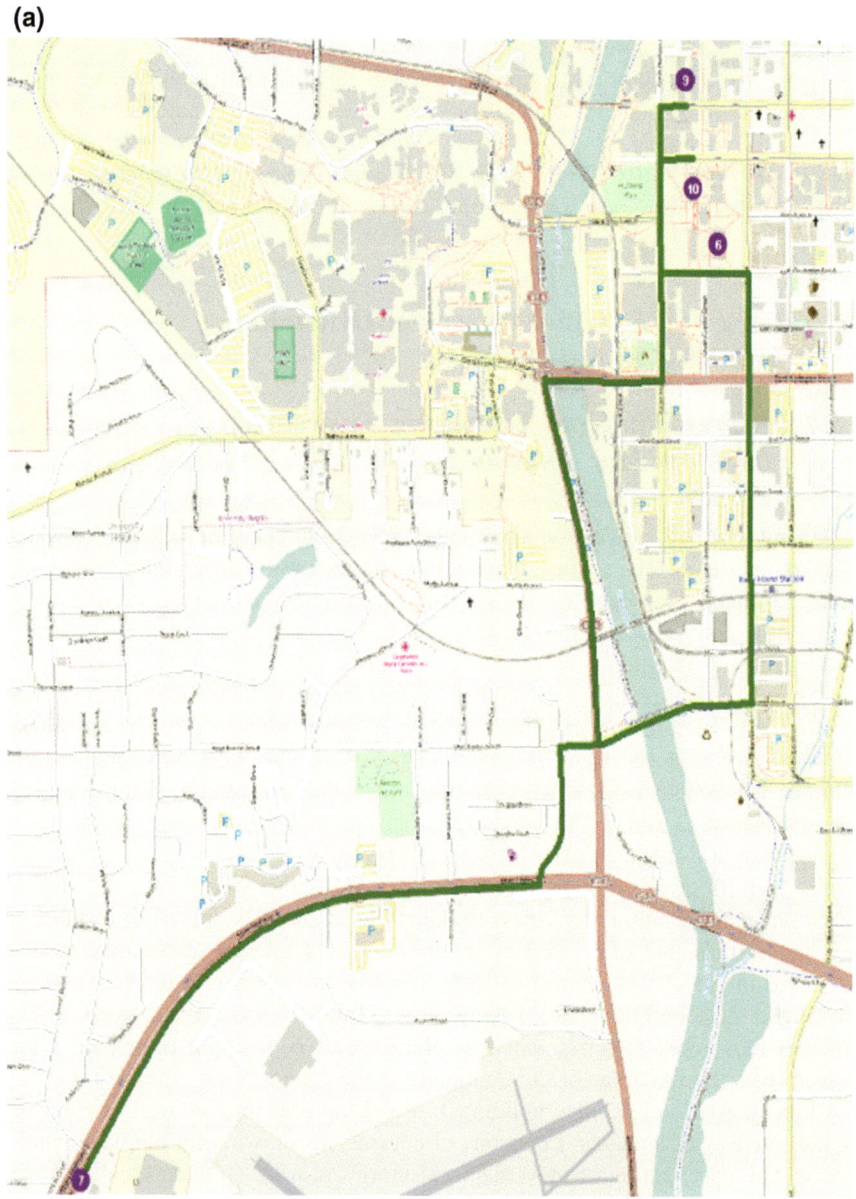

Fig. 5.5 Weekly movement trajectories for **a** student A, and **b** student B

It should be noted that although we generate a continuous path from the scheduled information, there may actually be time periods or *gaps* between courses such that the start of the next segment of movement in the trajectory may not begin

(b)

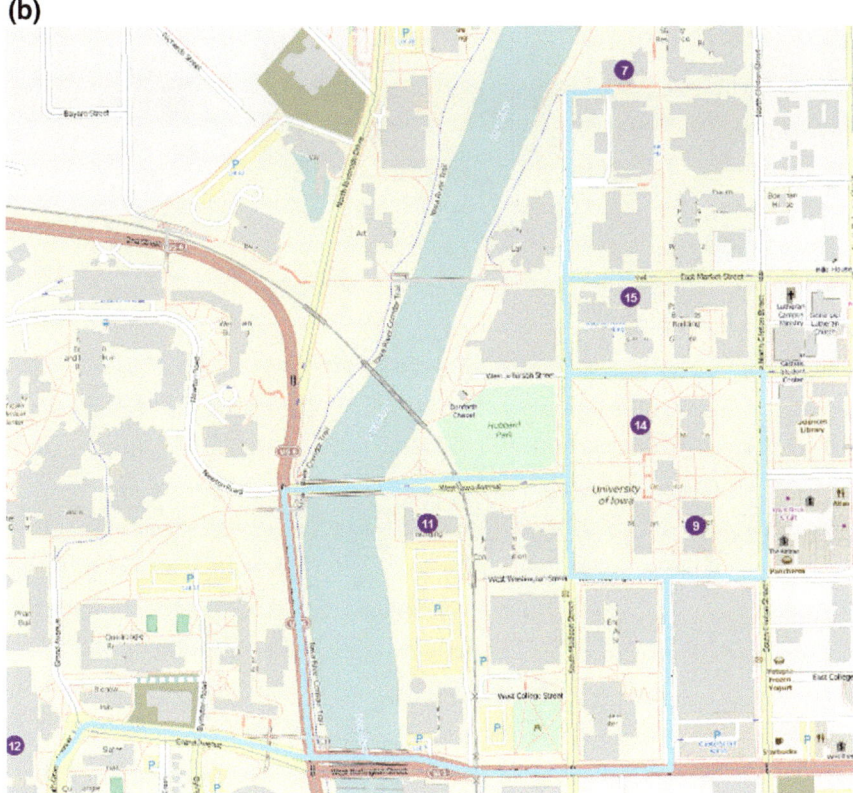

Fig. 5.5 (continued)

until some time has passed and it is time for the student to move on towards their next class. Another caveat of adopting road networks into the analysis is that the road network space overrules other possible movement paths, and provides one possible trajectory out of a set of possible trajectories. For example, the movement trajectory for student *B* (Fig. 5.5b) based on the 5 scheduled courses on the east side of the campus may not always hold as it is likely that students don't always use the road network and may cut across green spaces or pedestrian areas or take advantage of any available shortcuts. We show the spatiotemporal patterns of movement on campus using Euclidean space in Fig. 5.6. This choice yields a different yet still valid pattern of students' movement. The collective trajectories of the students for their Monday classes (Fig. 5.6a) are similar to that on Wednesday (Fig. 5.6b); and Tuesday and Thursday movements also share similarities. Students can have very different trajectories around campus and spend time in different locations,

(a)

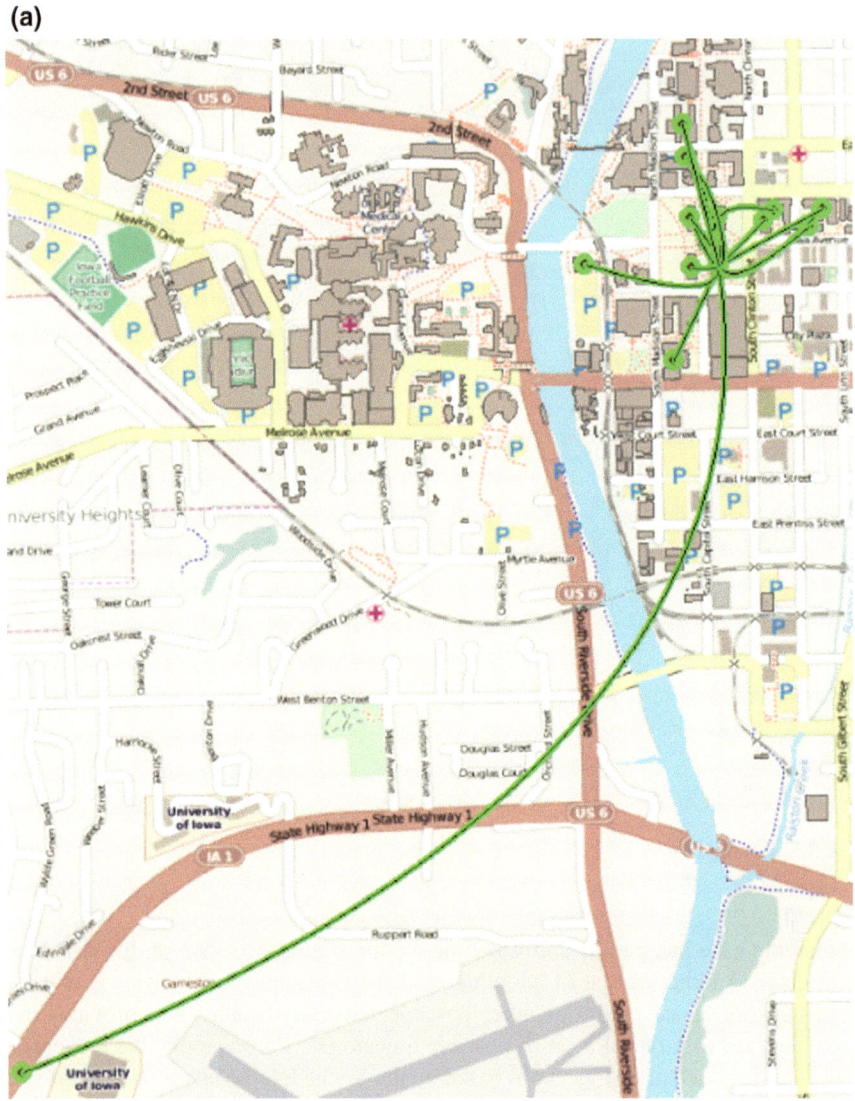

Fig. 5.6 Collective weekly trajectories of 13 students: **a** trajectory for monday, **b** trajectory for wednesday, **c** trajectory for Friday, and **d** trajectories for a whole week

the shortest weekly travelling distance out of this group of students is approximately. 5 miles indicating a schedule where classes are spatially very close, while the longest distance travelled is approximately 8.5 miles.

(b)

Fig. 5.6 (continued)

5.5.2 Spatiotemporal Patterns of Student Movment Paths

Representing collective trajectories at different temporal scales provides interesting insights regarding human movement dynamics. For example, the collective

(c)

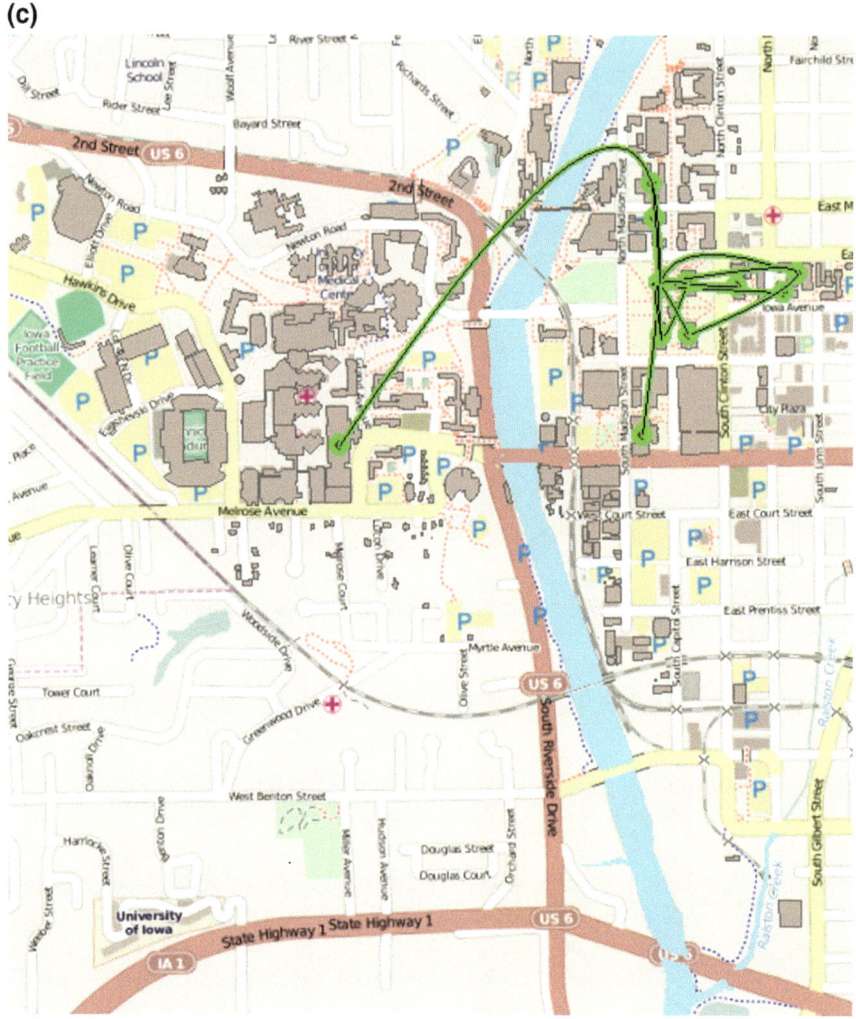

Fig. 5.6 (continued)

movement paths of all the students for the whole week (Fig. 5.6d), revealing that most of the travel for this group of students is in the area involving four buildings on the center of UIowa campus. Such a tool is helpful for security officers who can identify the busiest area on campus, i.e., areas that have the largest numbers of students moving around. The ability to identify such areas can help manage campus security and safety more effectively. From an academic standpoint, universities can utilize students movement information to help them arrange the space and time of courses more intelligently and direct resources, e.g., snow and ice removal in winter on highly accessed areas.

(d)

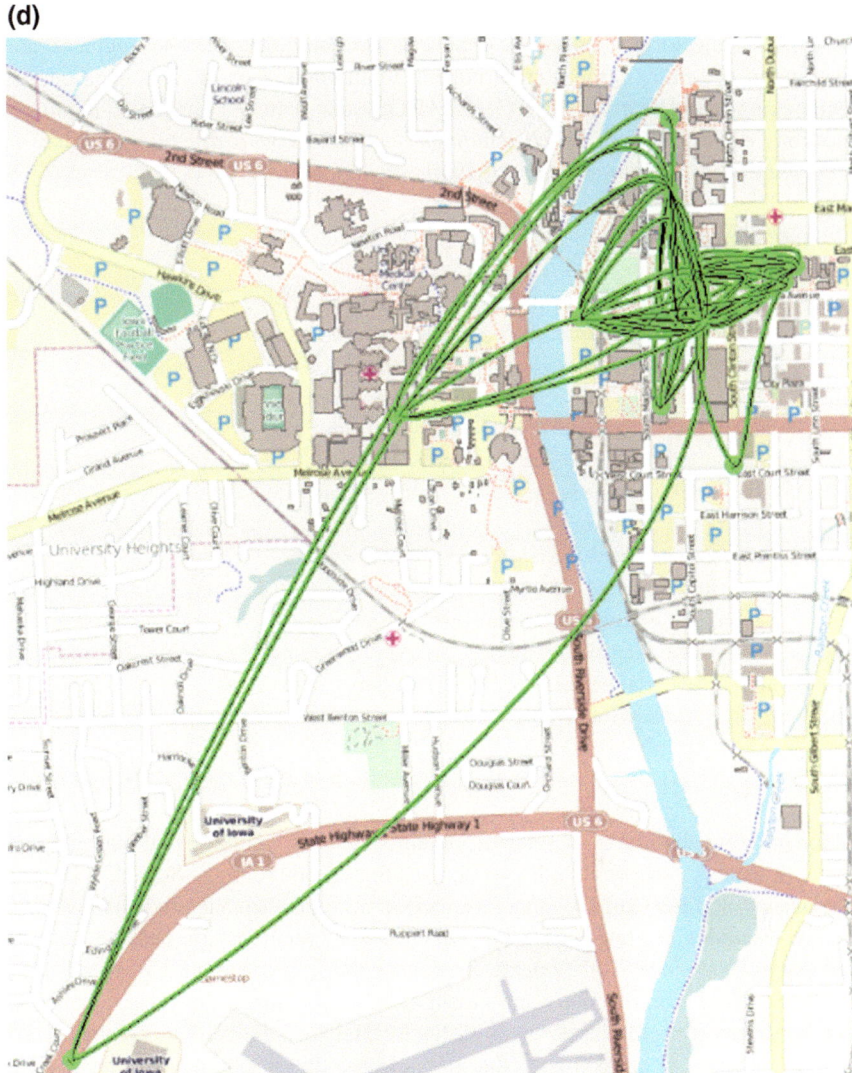

Fig. 5.6 (continued)

5.5.3   Tracking the Flow of Students Among Buildings

The linked nature of geospatial RDF knowledgebase offers us the ability to query and visualize the chain of movements of students on campus. Information about the inflow and outflow of students for a building at a specific time and modeling movement trajectories can offer important opportunities for management and planning of the entire campus environment. In addition, a set of movement queries become possible. For example, a GeoSPARQL query can be constructed to answer questions such as: *From which buildings do students who are taking classes in Schaeffer Hall on Wednesday at 11:30 am come from?*

```
SELECT ?pcourse
WHERE {
        ?course      rdf:type              : st_activity.
        ?course      :hasPhysicalSpace           ?building.
        ?building    :hasName            "Shaeffer Hall".
        ?course      :hasDayofWeek      time:Wednesday.
        ?course      :hasParticipant     ?student.
        ?student     rdf:type      foaf:agent.
        ?student     :hasTrajectory         ?traj.
        ?course      :partOf        ?traj.
        ?course      :previousActivity      ?pcourse.
        ?pcourse     rdf:type        :st_activity.
        ?course    time:hasBeginning  ?bgt.
        ?course    time:hasEnd          ?edt.
    FILTER     (xsd:datetime(?bgt) < "11:30am"^^xsd:time
                &&xsd:datetime(?edt) > "11:30am"^^xsd:time) .
    }
```

By utilizing the chain of semantic relations among courses, building and students, this query retrieves the list of courses that are ongoing in Schaeffer Hall on Wednesday at 11:30 am first (*hasPhysicalSpace*), and then the students and their course schedules are retrieved using *hasTrajectory* relationships between *Student* and *Trajectory*. Since all the course activities within the same individual trajectory are linked via *previousActivity* and *nextActivity* relationships, the classes that these students were in before arriving at Schaefer Hall can be retrieved. Note that a temporal interval is applied such that only classes that are scheduled on the same day (i.e., Wednesday) will be retrieved during the reasoning process. The visualization is generated based on the 13 schedules we collected, and for this reason, reflects a partial picture of the inflow movement to that building (Fig. 5.7). The complete pattern of inflow movements for Shaeffer Hall can be visualized provided schedules for all students are available.

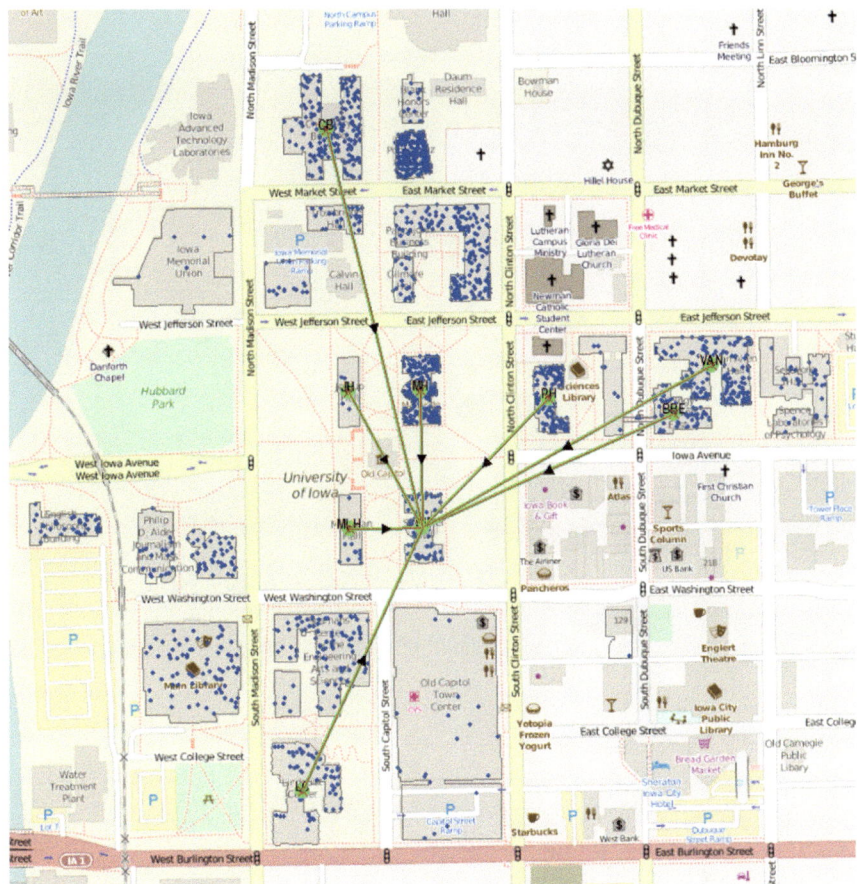

Fig. 5.7 Inflow of students to Schaeffer Hall on wednesday at 11:30 am

5.5.4 Spatiotemporal Aggregated Dynamics on Campus

Knowing about the ongoing courses at any time and the number of students in a building will be very useful for space and scheduling administrators, for example, as part of a smart campus effort, e.g., creating a smarter plan for energy consumption of the building or for supplying location-based services. This GeoSPARQL query can be extended to query all the buildings on campus at different times. The results of the extended queries yield an aggregated view of campus dynamics (Fig. 5.8). In Fig. 5.8, two snapshots show buildings on the east side of the campus on a Wednesday at 11:15 am and 7:15 pm. Each blue dot inside the building represents five students. The number of students at each time is

(a)

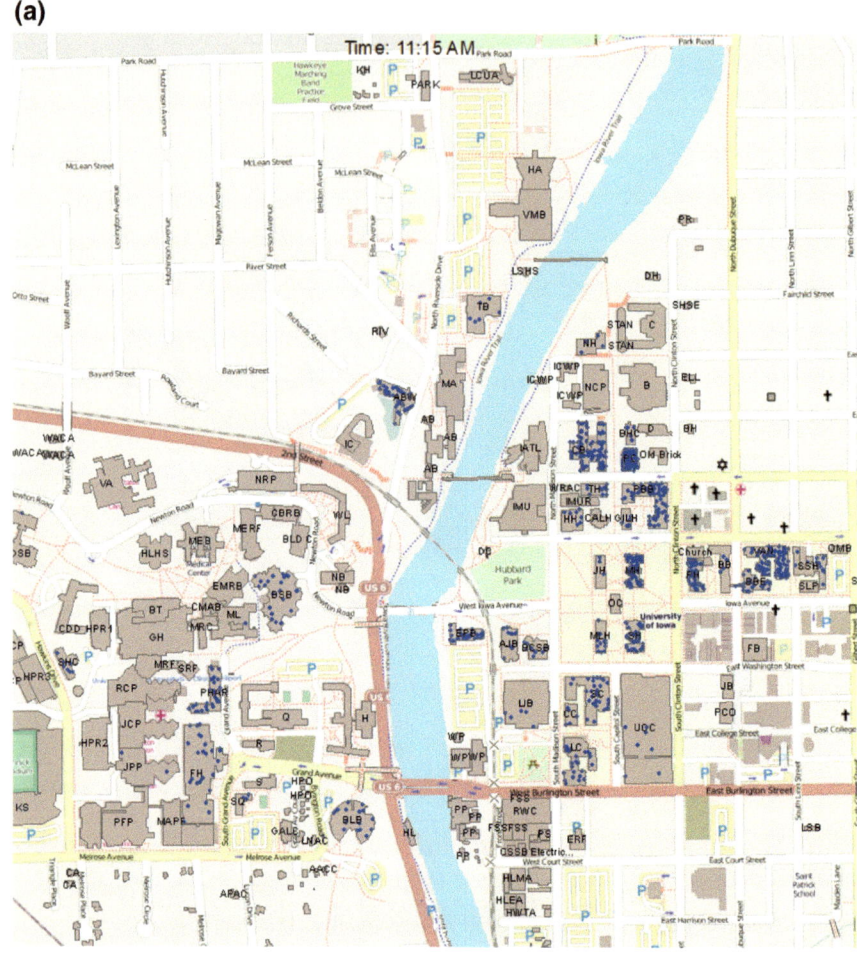

Fig. 5.8 Campus dynamics for a wednesday during fall 2014: **a** at 11:15 am, and **b** at 07:15 pm

calculated based on the enrollment number of each course. From the visualizations of these two different times, the diffusion of students across campus are revealed. Figure 5.8a is the snapshot at 11:15 am, where the number of students is beginning to reach a high level (262 courses with approximately 8000 students). After 5:30 pm, the number of students is starting to decrease such that by 7:15 pm there are about 130 courses going on with approximately 3300 students attending (Fig. 5.8b).

(b)

Fig. 5.8 (continued)

5.6 Conclusions and Future Work

This chapter presents an approach that uses semantic web technologies for representing, querying, reasoning, and visualizing about movements associated with scheduled events–in this case, courses and seminars–on a university campus. This framework provides a method for integrating data from spatial, temporal and semantic dimensions, where semi-structured text information about scheduled courses can be transformed into a structured RDF data model that is further enriched with spatial and temporal data. This work demonstrates the insights to be gained from incorporating spatial thinking into facilities management and planning, creating physical environments that are context-sensitive and smart. Representing spatial information associated with scheduled activities provides a basis for modeling the movements of students in response to scheduled course activities at different temporal granularities. Individual schedules can be visualized at the granularity of daily or weekly personal trajectories. Alternatively, aggregate views showing where students are clustered at different locations and times capture movements at a different granularity. Visualizations of scheduled activity spaces show how students and faculty use campus space and the move about on campus. For campus management, management staff gain visualizations of the spatiotemporal dynamics for the entire campus, providing better insights into facilities usage as well resource (e.g., energy) consumption. Furthermore, semantically enriched events (e.g., spatially fixed temporally flexible) can give users the ability to reason using the ontology and provide cues for intelligently adjusting daily schedules depending on the location information. This ontology-based framework can be adapted to other application domains where individuals undertake scheduled activities, for example, transportation and logistics planning where aggregated dynamics about individual movements (e.g., pedestrian and vehicular traffic) are also important. A future extension for this framework is to further distinguish between fixed and flexible activities. The reasoning engine can be extended to reason about the spatiotemporal interactions between different types of space-time activities, and derive spatial prisms for individuals based on his/her scheduled activities.

Appropriately representing, organizing, and processing spatiotemporal data is a fundamental research theme in GIScience community, especially as place-based research is evolving into people-based research (Miller 2007). Modeling activity-related information is critical for a better understanding human behavior and movement pattern. Using semantic web technologies to model space-time activities allows researchers to link together diverse data sources related with human mobility based on ontology matching. This is more flexible than relational database model both in terms of data integration and data analysis as the triple pattern based query allows users to express semantics in a more intuitive manner. Geospatial RDF data are becoming more and more prevalent, including OpenStreetMap and GeoNames etc., providing opportunities for modeling the context of human mobility behavior more realistically. Future work could consider more realistic representation of context including environmental data, social data, additional real-time events, and even social network information of individuals. Incorporating these contextual variables that influence human behaviors and mobility will improve movement and activity space modeling in the information age.

References

Abukhater, A., & Walker, D. (2010, July). Making smart growth smarter with GeoDesign. *Directions Magazine.*

Afyouni, I., Ilarri, S., Ray, C., & Claramunt, C. (2013). Context-aware modelling of continuous location-dependent queries in indoor environments. *Journal of Ambient Intelligence and Smart Environments, 5*(1), 65–88.

Allen, J. F. (1984). Towards a general theory of action and time. *Artificial Intelligence, 23*(2), 123–154.

Andrienko, G., Andrienko, N., Bak, P., Keim, D., Kisilevich, S., & Wrobel, S. (2011). A conceptual framework and taxonomy of techniques for analyzing movement. *Journal of Visual Languages & Computing, 22*(3), 213–232.

Andrienko, G., Andrienko, N., Hurter, C., Rinzivillo, S., & Wrobel, S. (2013a). Scalable analysis of movement data for extracting and exploring significant places. *IEEE Transactions on Visualization and Computer Graphics, 19*(7), 1078–1094.

Andrienko, N., Andrienko, G., & Fuchs, G. (2013b). Towards privacy-preserving semantic mobility analysis. In *In EuroVis workshop on visual analytics. The Eurographics Association* (pp. 19–23).

Calderoni, L., Maio, D., & Rovis, S. (2014). Deploying a network of smart cameras for traffic monitoring on a 'city kernel'. *Expert Systems with Applications, 41*(2), 502–507.

Chen, J., Shaw, S.-L., Yu, H., Lu, F., Chai, Y., & Jia, Q. (2011). Exploratory data analysis of activity diary data: A space-time GIS approach. *Journal of Transport Geography, 19*(3), 394–404.

Crease, P., & Reichenbacher, T. (2013). Linking time geography and activity theory to support the activities of mobile information seekers. *Transactions in GIS, 17*(4), 507–525.

Crooks, A., Croitoru, A., Stefanidis, A., & Radzikowski, J. (2013). #Earthquake: Twitter as a distributed sensor system. *Transactions in GIS, 17*(1), 124–147.

Egenhofer, M. J. M. (2002). Toward the semantic geospatial web. In *Proceedings of the tenth ACM international symposium on Advances in geographic information systems—GIS'02* (pp. 1–4). New York, New York, USA: ACM Press.

Galache, J. A., Sotres, P., Santana, J. R., Gutierrez, V., Sanchez, L., & Munoz, L. (2013). A living smart city: Dynamically changing nodes behavior through over the air programming. In *Proceedings—27th International Conference on Advanced Information Networking and Applications Workshops, WAINA 2013* (pp. 1271–1276).

Giannotti, F., Nanni, M., Pedreschi, D., Pinelli, F., Renso, C., Rinzivillo, S., et al. (2011). Unveiling the complexity of human mobility by querying and mining massive trajectory data. *The VLDB Journal, 20*(5), 695–719.

Goodchild, M. F. (2007, November). Citizens as sensors: The world of volunteered geography. *GeoJournal, 69*, 211–221.

Grenon, P., & Smith, B. (2004). SNAP and SPAN: Towards dynamic spatial ontology. *Spatial cognition and computation, 1* (March), 69–103.

Hägerstrand, T. (1970). What about people in regional science? *Papers of the Regional Science Association, 24*(1), 6–21.

Janowicz, K., Scheider, S., Pehle, T., & Hart, G. (2012). Geospatial semantics and linked spatiotemporal data—Past, present, and future. *Semantic Web, 3*(4), 321–332.

Kuhn, W. (2005). Geospatial semantics: Why, of what, and how? In S. Spaccapietra & E. Zimányi (Eds) *Journal on Data Semantics III. Lecture Notes in Computer Science* (Vol. 3534). Berlin, Heidelberg: Springer.

Kuhn, W. (2012). Core concepts of spatial information for transdisciplinary research. *International Journal of Geographical Information Science, 26*(12), 2267–2276.

Kwan, M.-P., Janelle, D. G., & Goodchild, M. F. (2003). Accessibility in space and time: A theme in spatially integrated social science. *Journal of Geographical Systems, 5*(1), 1–3.

Majid, A., Chen, L., Chen, G., Mirza, H. T., Hussain, I., & Woodward, J. (2013). A context-aware personalized travel recommendation system based on geotagged social media data mining. *International Journal of Geographical Information Science, 27*(4), 662–684.

Miller, H. J. (1991). Modelling accessibility using space-time prism concepts within geographical information systems. *International Journal of Geographical Information Systems, 5*(3), 287–301.

Miller, H. J. (1999). Measuring space-time accessibility benefits within transportation networks: Basic theory and computational procedures. *Geographical Analysis, 31*(2), 187–212.

Miller, H. J. (2005). A measurement theory for time geography. *Geographical Analysis, 37*(1), 17–45.

Miller, H. J. (2007). Place-based versus people-based geographic information science. *Geography Compass, 1,* 503–535.

Miller, H. J. (2014). Activity-based analysis. In M. M. Fischer, & P. Nijkamp (Eds.), *Handbook of Regional Science* (pp. 741–758). Berlin, Heidelberg: Springer Berlin Heidelberg.

Miller, H. J., & Goodchild, M. F. (2014). Data-driven geography. *GeoJournal.*

Neutens, T., Van de Weghe, N., Witlox, F., & De Maeyer, P. (2008). A three-dimensional network-based space-time prism. *Journal of Geographical Systems, 10*(1), 89–107.

Perry, M., Sheth, A. A. P., Hakimpour, F., & Jain, P. (2007). Supporting complex thematic, spatial and temporal queries over semantic web data. *GeoSpatial Semantics,* 228–246.

Perry M. S. (2008). *A framework to support spatial, temporal and thematic analytics over semantic web data.* Wright State University.

Roche, S. (2014). Geographic information science I: Why does a smart city need to be spatially enabled? *Progress in Human Geography, 38*(5), 703–711.

Sengupta, S., Ganeshan, K. V. V., & Sarda, N. L. (2010). Developing IITB smart campusGIS grid. In *Proceedings of the 1st Amrita ACM-W Celebration on Women in Computing in India—A2CWiC'10* (pp. 1–8). New York, New York, USA: ACM Press.

Shaw, S.-L., & Yu, H. (2009). A GIS-based time-geographic approach of studying individual activities and interactions in a hybrid physical–virtual space. *Journal of Transport Geography, 17*(2), 141–149.

Smith, B., & Grenon, P. (2004). The cornucopia of formal-ontological relations. *Dialectica, 58*(4), 279–296.

Sowa, J. F. (1999). *Knowledge representation: Logical, philosophical and computational foundations.* Brooks/Cole Publishing Co.

Stewart, K., Fan, J., & White, E. (2013). Thinking about space-time connections: Spatiotemporal scheduling of individual activities. *Transactions in GIS,* 791–807.

Wang, M., & Ng, J. W. P. (2012). Intelligent mobile cloud education: Smart anytime-anywhere learning for the next generation campus environment. In *2012 Eighth International Conference on Intelligent Environments* (pp. 149–156). IEEE.

Yin, L., & Shaw, S.-L. (2015, September). Exploring space-time paths in physical and social closeness spaces: a space-time GIS approach. *International Journal of Geographical Information Science,* 1–20.

Yu, H. (2006). Spatio-temporal GIS design for exploring interactions of human activities. *Cartography and Geographic Information Science, 33*(1), 3–19.

Author Biographies

Junchuan Fan (M.C.S., Ph.D., University of Iowa), Postdoctoral fellow at Center for Geospatial Information Science, University of Maryland. His primary research interests are spatiotemporal data modeling, geosocial computing and geospatial semantics. His research regarding human dynamics focus on the interactions between place semantics and human activity and mobility pattern. Specifically, he integrates semantic modeling framework with big geospatial data mining technologies to learn geographic knowledge about places and uncover the link between human activity spaces, movement patterns and place semantics.

Kathleen Stewart (Ph.D., University of Maine), is director of the Center for Geospatial Information Science and associate professor in the Department of Geographical Sciences at the University of Maryland. Her research focuses on the intersection between geographic information science and geographic dynamics. She is interested in mobility and big geospatial data and how these apply to different application domains, e.g., public health and transportation. She is also interested in geospatial ontologies and their role for GIS. She is currently a member of the Mapping Science Committee of the National Academies of Sciences, Engineering and Medicine, and the Board of Directors for the University Consortium of Geographic Information Science.

Chapter 6
Smart Sensors, Cyborgs, and Cybernetics: A Critical Reading of Smart City Technologies

Chen Xu

6.1 A World Replete with Digital Traces

A significant happening about modern society is the ever-increasing mobility that people are able to move both material objects and immaterial information across space. The accelerating pace of evolution of digital technologies to facilitate rearrangement of time and space has transformed human behaviors to an extent that Mitchell (2002) argued that there are two fundamental building blocks of this reality, namely bits and atoms. Modern information computation based technologies consume and produce immaterial substance consisting of 0's and 1's, the bit. Human bodies and natural or man-made environment are materials made up of atoms. Space-time is increasingly compressed as the socioeconomic consequence. Harvey (1990: 284) asserts that this dramatic process "had a disorienting and disruptive impact upon political-economic practices, the balance of class power, as well as upon cultural and social life." More empirical studies are thence needed as to scientifically understand the phenomenon. The demand for more empirical studies has been supported by the increasing availability of data generated by people using digital technologies. Recently, many studies have been conducted by examining digital traces left by social media users, which bring new discoveries about nature and society (Shaw et al. 2016). Techno-optimists generally believe that the availability of big amount of digital traces and many other automatic technologies would help create better cities, in other words, smart cities (Schaffers et al. 2011). But as technology in general lacks the ability to be self-initiative and is malleable to imposed intensions, a critical framework is equally important to examining the implications of technology.

C. Xu (✉)
Department of Geography, University of Wyoming,
1000 E University Ave, Laramie, WY 82071, USA
e-mail: cxu3@uwyo.edu

© Springer International Publishing AG, part of Springer Nature 2018
S.-L. Shaw and D. Sui (eds.), *Human Dynamics Research in Smart and Connected Communities*, Human Dynamics in Smart Cities, https://doi.org/10.1007/978-3-319-73247-3_6

Goodchild (2007) first delineates the phenomenon from a geographical perspective through the juxtaposition of spatial sensors and the general data creators—i.e., citizens-as-sensors, and theorizes the products as well as their features as volunteered geographic information (VGI). VGI denotes an ontological significance of user generated geographic data, which provides an alternative perspective about the Earth's surface beside the dominant authoritative view. Subjectivity is a crucial and valuable part of VGI, and the subject can assume many forms from human to things that only virtually exist. VGI can be by-products of digital technology involved activities, and these by-products act like digital traces revealing their producers' behaviors. Thence, the assembly of digital traces opens a window for speculating the corresponding producer's behaviors. Meanwhile, more people are using mobile devices to access the Internet and there are multiple means via mature technologies to locate a device (Cheng et al. 2010). Movement information of a subject becomes attainable through a proxy relationship between the subject and digital devices. The question then becomes whose movement is captured. Meanwhile, embedded in the subjective digital traces are people's proactive behaviors of receiving, filtering, and distributing information. People users' intelligence is augmented by technology and citizens-as-sensors are smart sensors. In this study, I focus on one specific type of digital traces, which are left by social media users. While there is a variety of social medias, Twitter has emerged because of its popularity worldwide and its relative openness for data collection.

Twitter is a micro-blogging service that claims to have 284 million monthly active users globally who post 500 million tweets daily and about 80% of them are using mobile devices to access the service.[1] Among the total amount of stored tweets, 1% of them are selected by Twitter's proprietary algorithm and published through several public application programming interfaces (APIs) in real-time, a mechanism called streaming. The streaming mechanism makes Twitter data attractive for researchers outside the company in academic fields from communication, psychology, to geography and etc., for the data provides a glimpse into social interactions of millions of people. Such an opportunity did not ever exist before the worldwide web evolved into 2.0 (Han 2011). Studies on tweets first center on Twitter as a social media platform, and topics range from examining information propagation pattern to networking connections between different users (Chechev and Georgiev 2012; Ye and Wu 2010). Recently, the subjective nature of tweets sees emergence of initial researches interested in exploring the intimate relationships between couples or the geographical awareness of individuals that are reflected in tweet contents (Garimella et al. 2014; Xu et al. 2013). With the rising of mobile computing, we see more tweets are posted from mobile devices that if location services are activated will create a trajectory record of the creator's situations. Thus, the agglomeration of continuous georeferenced tweets provides a window or a lens for observing people's spatiotemporal behavior. The construction

---

[1]According to Twitter company about information at https://about.twitter.com/company.

of behavioral patterns was leveraged in, e.g., urban studies (Huang and Wong 2016). The methodology is unobtrusive in terms of scientific data collection scheme, which means people would not know that their posts are collected. It provides global coverage as long as there are Twitter users in a geographical area. And compared to conventional survey methods, data collection via social media platforms is much more cost-effective; for once the essential computing infrastructure has been built, data can be archived without incurring much extra cost. However, the strong subjective nature of the data creates differences that need to be examined carefully. This chapter compares traditional methods for collecting daily life traces to alternative data-based approaches in human dynamic studies; then proposes a new identity as cyborg to synopsize the differences; the relations between cyborgs and their environments especially cities are examined by using a theoretical framework of cybernetics to argue some implications of technology-backed version of smart cities.

6.2   Daily Life Traces Collecting in Conventional Geographic Studies

The interweavement of individual humans' everyday life trajectories at the macro-level sketches the societal dynamics out in a dazzling yet reflective means (Ellegard 1999; Hägerstrand 1991). In a study about the dynamics of overlapping work and personal life, the researcher made in depth observations about the study subjects who suffered stress due to the unavoidable mixture of public and private lives (Renshaw 1976). Individual adults play multiple roles in contemporary society, and tensions if unattended can ramify from one organizational system to another through individuals as the medium. In another study, researchers spent eight months in an agricultural cooperative to observe community members acting in their daily personal lives and works, whose micro-level activities eventually led to decision-makings that influence the stability and the sustainability of communities at the local as well as the whole society in the macro level (Kroeker 1995). Meanwhile, individual expectations collectively influence the formation of societal organizations, which profoundly impact the economic performance and growth of a society (Greif 1994). Thus, the depictions of individuals and their daily lives are crucial for understanding societies, as societies consist of various social systems that are artifacts resulting out of human designs (Buckley 1998).

The pivotal position of individuals performing daily activities in producing societal dynamics also signifies the importance of space and time in this process, for space and time are fundamental dimensions of human life (Kellerman 1989). Geographers have a long tradition in deciphering human societies using the fabric of space and time as the framework to untangle the perplexed thread of human behaviors, for example, as their intensions to carve out meaningful places in the space-time continuum for anchoring personal identities and experiences (Tuan 1977;

Buttimer 1976). After Hägerstrand's abstraction of a space-time prism construct and upon which the formation of a time-geographic framework, geographers are more capable of quantifying the effects of both the interactions between and the constraints of the geographic space and time that would eventually lead to the presence or absence of opportunities for individuals (Miller 1991, 1999). In Kwan's (1999) study, a two-day travel diary dataset of 56 full-time employed European Americans was collected to assess gender related differences in accessing to opportunities. In a study to estimate the effect of distance on individual's accessibility to job opportunities, Weber (2003) used a subset consisting of 775 adults' travel-activity diaries, which were extracted from a larger dataset of a survey conducted by the Portland Metropolitan Service District that surveyed 10,084 individuals. Wong and Shaw (2011) proposed an activity space approach to measure segregation and used a dataset comprising travel diaries contributed by about 5000 households to illustrate their approach. Despite differences in methodologies and variations in research foci, the aforementioned studies shared some somewhat close conclusions which point out the relationships between individuals and opportunities by various social systems are complex and cannot be easily decoded by merely measuring geographic distance or by solely examining the residential space; meanwhile, findings are hard to be generalized due to the intricate nature of human subjects. Personal experiences, backgrounds and habits play significant roles in formulating personal opportunities to a great extent that in some cases even reversing our conventional understandings. Thus, the necessity of continuing explorations is warranted.

Travel diary methods or time diary methods are commonly implemented in studies concerning individual activity patterns by recording the geographic space-time footprints of human subjects (Axhausen et al. 2002; Janelle et al. 1988). Travel diaries capture activities that would be categorized into different activity classes, and normally such data are companied by surveys on the socioeconomic statuses of the subjects to provide the necessary personal contexts. Although they are commonly implemented in a wide spectrum of applications from transportation planning to tourism research, the diary based methods put heavy burdens on the study subjects; the effectiveness of the methods as well as the validity of results rely largely on the human agents making commitment to faithfully and accurately record location and time information (Wolf et al. 2001). Especially when the duration of a survey is to cover a longer period of time other than just a few days, the complexity of management becomes even more prominent (Axhausen et al. 2002). The commercialization of global positioning systems and their implementations in travel data collection automate the data collection procedure and relief the participants from meticulous input operations; consequently, data quality has been greatly improved through eliminating man-made errors and new types of data can be collected such as instantaneous velocity and travel route (Wolf et al. 2001). Meanwhile, data collection becomes more cost-effective (Stopher et al. 2002). Since then, with the emerging of ICTs, means for human activity data generation has become even more versatile (Soora 2014; Oliver et al. 2010; Kwan 2007).

In general, the availability of new technologies empowers researchers toward the augmented capabilities to examine human behaviors from increased numbers of aspects (Draijer et al. 2000). New challenges emerge as well, for example, due to limitations of machines to pickup contextual cues of human activities, comparably there is less amount of information in automated diaries; also, data quality is greatly influenced by the availability and adequacy of positioning technologies, e.g., the number of satellites or the coverage of information communication networks; and while survey participants are relieved from constant data logging, they are now charged with responsibilities to maintain the operability of their electronic tracking devices which generally have a battery component (Oliver et al. 2010; Wolf et al. 2001). Meanwhile, the discussion of using new positioning technologies to auto-mate data collection has an indispensable prerequisite, the availability of funda-mental infrastructure such as information communication networks or the GPS networks. Thus, the methods would be far less useful in developing countries, which have less than adequate availability of ICT infrastructure or positioning systems (Kwan 2007).

Nonetheless, the advancements in technologies as well as in research method-ologies still vastly push forward our understandings about human behaviors in their daily lives (Wiehe et al. 2008). Among all the technologies, geographic information system (GIS) has made unique contributions by offering the functionalities for plotting behavioral datasets against the spatiotemporal dimensions. Hence it creates a proxy of human daily life out of the abstract spatiotemporal data and gives the data a comprehensible geographic context, and more crucially it provides a framework for quantification (Shaw and Wang 2000). From the visualization per-spective, Hägerstrand's space-time prism is still by far the most prominent framework, which consists of a geographic plane (x- and y-axis) and a temporal dimension (z-axis), for space-time paths or trajectories (Kraak 2003). The com-plexity of human behaviors renders it difficult to rely exclusively on machines to filter out patterns beyond mechanic interpretation. A synergistic collaboration between machines and human visual system is far more capable and effective to identify anomaly out of norm, such an approach has been termed visual analytics (Andrienko et al. 2010). Thus, the implementation of automated technologies starts to peel away context from the geographic data. Extra efforts are required for the contextual information acquisition; at this moment it is still achievable for people to synthesize these two parts through creative analysis.

A lack of progress in time geography research after Hägerstrand can somehow attribute to the scarcity of spatiotemporal data that reflect human lives as well (Kwan 2004). Rapid advancements in computational technologies have quickly changed the field of geography from, like the ongoing transformation in many other fields, a situation in constant need of data and computation resource to a situation of data abundance and computation resource affluence (Miller and Han 2009). Progresses in ICTs, positioning technologies and social media contribute to the vast expansion in data collections—a phenomenon called Big Data in general and data avalanche in geography (Sui and Goodchild 2011; Miller 2010). In the fields of geography, especially in the field of human geography, the sudden availability of big amount of spatiotemporal data, which are digital traces left by human activities in the hybrid space, offers opportunities to push forward geographical understanding (Kitchin 2013). Outside geography, researchers in other fields already started to grasp the opportunities to use big data to study human society at scales that were difficult to look at before (Cranshaw et al. 2012; Noulas et al. 2011). Geographers also have been eager to leverage the data contributed by volunteers as well as left by social media users to experiment alternative solutions for challenges facing geography (Zook et al. 2010; Elwood 2009).

6.3 The Emerging Alternative Ways for Activity Traces Collection

Understanding dynamics of human mobility via massive amount of mobile data is potentially beneficial for effective management of societies for tasks from urban planning, transportation management to prediction of infectious disease outbreak (Shaw et al. 2016; Laurila et al. 2012; Phithakkitnukoon et al. 2010; González et al. 2008). However, such datasets have serious privacy issues that their usages are highly regulated by laws (Minch 2004). Also, historically only mobile phone companies, like more recently web service companies, have access to the entire datasets, which were rarely opened to outside researchers. In order to get hold of a part of such valuable mobile data, several strategies were implemented. Laurila et al. (2012) allocated cellphones with data collection software installed to a group of about 200 volunteers in Switzerland. They were able to collect locational, temporal, velocity, as well as communication content information about the volunteers. Phithakkitnukoon et al. (2010) received mobile phone data from a third party, the dataset include about one million records of users in Boston metropolitan area. González et al. (2008) used anonymized billing dataset by a mobile phone service provider of about 100,000 users. The dataset provided positions of the users for a period of six months. The broad user base together with abundant mobile applications has made mobile phones a popular sensing platform. These pioneer studies have made significant discoveries about human mobility pattern in both space and time dimensions. One of the many typical discoveries by using mobile phone data is that the better revealing of daily rhythmic patterns of human society across different geographical areas (Noulas et al. 2011).

The emergence of mobile social media applications immediately provides researchers an alternative means for study human dynamics. Many such studies are based on data harvested from Twitter, because of the relatively open data policy of the platform. Although many social media data are generated by applications on mobile phones, dataset from the same physical platform shows quite different traits. Mobile phone data are created passively whenever the phone makes a connection with a station; thus, although there is close bond between the phones and their users, human factors have little influence in the data life cycle (Ahas et al. 2007).

Social media data, on the other hand, is human's active creation. And even for data from a simple social media service like Foursquare which provides service for friends to connect visited places, Noulas et al. (2011) points out the human factors in the data analysis, because whether or not to post a place-based entry is entirely depends on the control of the person. In general, social media data is strongly subjective in nature (Hogan 2010). Thus, social media contents, in the case of this study tweets, exhibit personal characteristics, such as political bias or place preference (Xu et al. 2013; An et al. 2012). Adding to this difficulty is that the process leading to the selection of tweets to be in the public data stream is uncontrolled by and worse opaque to the data users or the researchers (Urquhart and Vaast 2012).

As long as social media companies put restrictions on their data access and keep that algorithm-based data sampling policy hidden, there would not be a thorough understanding about the nature of any arbitrarily collected dataset. However, even we only have accessed to the 1% of public real-time tweets, the agglomeration of data through time can be enormous in size, which gives the impression of the phenomenon called Big Data for many people. However, in Big Data a larger data size cannot be equal to a deeper insight or richer information about the population (McNeely and Hahm 2014). Thence, the using of social media data for studying human mobility is not only methodologically but also epistemologically challenged (Boyd and Crawford 2012). Consequently, meaningful social media data facilitated studies would start with the construction of identities of data producers. Special attention needs to be paid for the interpretation of identity. Identity in cyberspace has higher complexity than it in physical space because of the online ambiguity (Kendall 1998). For online and offline lives increasingly become two sides of the same coin, the subject behind the embodied identity in digital traces can be the atom, the bit, or the synthesis of both, because people with the augmentation of mobile devices now are able to act in both cyberspace and physical space simultaneously (Jurgenson 2012). In other words, there is a mixture of corporeal mobility and information mobility in the integral spatial mobility that can be discovered from the collected Twitter data. Xu et al. (2015) presents examples that show the two types of spatial mobility from the same data source. The authors used a space-time cube approach to illustrate four different activity trajectories of Twitter users. Apparently as information takes no time to cross physical distance, information's mobility shows much higher mobility than corporeal's. However, this difference in subject identity is not immediately clear by merely reading the trajectories.

Using digital traces left by mobile device users to study human behaviors has been carried out since anonymized cellphone data was occasionally available for researchers to grab (Isaacman et al. 2011). Cellphones have to contact cellular towers for services and such activities left records at the servers. Because the data is sensitive to users' privacy, therefore the records have to be anonymized before publishing. Comparing to cellphone data, social media records share several similarities. First, their production is triggered by individuals' communication activities. Second, they record both location and time information about such activities. And third, they are mobile, which means they can be treated as proxy for places people have been to. But, indeed, social media records are far more different from cellphone data. First, in cellphone data, the location is accurate to the grid of base stations, since the cellphone only reports the location of the cellular tower that it is contacted. Compare to cellphone based approach location accuracy of social media records has a wide variation depending on the means implemented for location information. Second, while the identity of individuals when decoding cellphone data is homogeneously human subjects, the identities reflected in social media data have to be examined more closely and carefully. Third, regarding the content richness, social media records offer much richer information about their users. For example, in the case of Twitter, every published entry may include profile contents about the account, information about the receiver(s), and more importantly the textual content.

6.4   Cyborg—Is It the New Identity?

As the fast advancement of wearable technologies and the increasingly matured implant technologies, communicative machines will play unprecedentedly critical roles in human's daily lives (McGee and Maguire 2007). Then there is the rising of manmade things that can be connected to each other, the so-called Internet of Things (IoTs). Communication in terms of transmitting and exchanging information is built into these things, which gives a various things the capacities to communicate with humans or each other. During Hurricane Julio 2014, a robotic floating drone was deployed around Hawaii. People could request oceanic climate status by sending demands through Twitter. For example, sending "@Holohoho_wg where are you?" could in return have the geographic location of the drone.[2] Meanwhile, the drifting on ocean robot sent tweets of photos or texts automatically, which could be collected through one of the Twitter's application program interfaces (APIs). For example, Fig. 6.1 shows a communication activity happened between the robot and a human Twitter user.

Data of this event was published via the API in the format of JavaScript Object Notation (JSON). The content, time and geographic location about the event then can be compiled by extracting correspondent information from the key-value pair. For the above example, the data provides the following attributes.

'text': '"#robotics #drone @ here's the picture you requested! https://t.co/1kDODWcmRU"',

'coordinates': {'type': 'Point', 'coordinates': [−155.89324583, 20.01580317]},

'created_at': 'Tue Aug 30 00:37:21 +0000 2016'

The assembly of these data thus provides a spatial trajectory of the subject about its movement in the physical world. By far the most popular Twitter users are human subjects and the tweets they left behind containing the aforementioned types of

---

[2]https://twitter.com/Holoholo_WG.

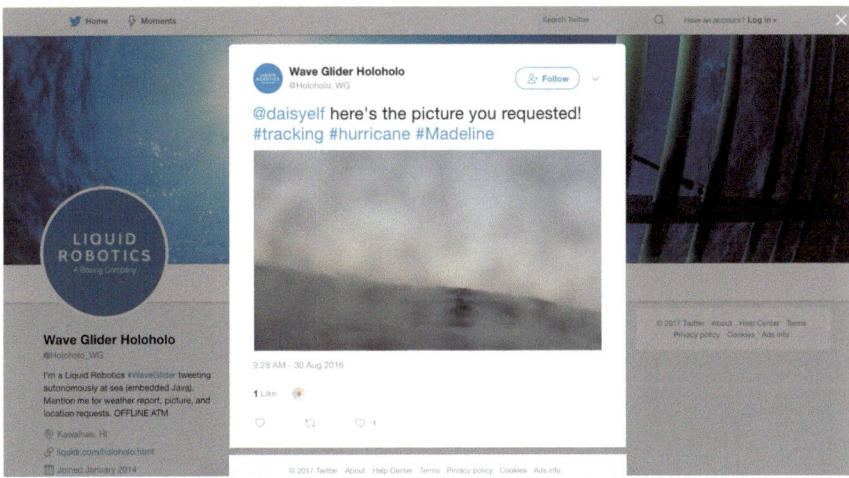

Fig. 6.1 An example of information exchange between a human Twitter user and a robot Twitter user https://twitter.com/Holoholo_WG/status/770659561564803072/photo/1

spatiotemporal information thence reflect Twitter user activities. Figure 6.2 presents two different trajectories by plotting geographic locations along the temporal dimension as when a human Twitter user posted a tweet at that location.

The authors indicated that the data was collected between February 7th and 20th, 2015. A node on the map represents the location of a tweet, where a Twitter user made a post. An edge links two event locations which when displayed in a space-time cube give the temporal sequence. Within the same period of time, apparently the activity intensities of the two users are dramatically different. It calls into question the velocity and intensity of movements in the right column that could be achieved by a human being. After collecting the data by the two users and examining the contents closely, two different types of user identities become clear. While the trajectory on the left side represents a real person's activity path, the trajectory on the right side belongs to a group of real persons sharing a Twitter account. If a real person's historical tweets can be used for studying corporeal mobility, it is erroneous to make the same conclusions for a group of persons.

Indeed, what the trajectory on the right side reveals is the movement of information among a group of Twitter users, who post tweets to the same user account. Therefore, it might be stated that the information exchange records inside the group can be used for studying information's spatial footprint. Members at different geographic locations share any messages posted to the account; and the receivers might then distribute the message to their friends dispersed at different locations. As information takes no time to cross long distance, the vehement velocity represented by the right side trajectory becomes reasonable. The great impact of information being able to fiercely reconfigure space-time relations cannot be overstated. Ample amount of examples can be enumerated to shed light on the issue. It would be an impossible mission for OpenStreetMap (OSM) volunteers located globally to help

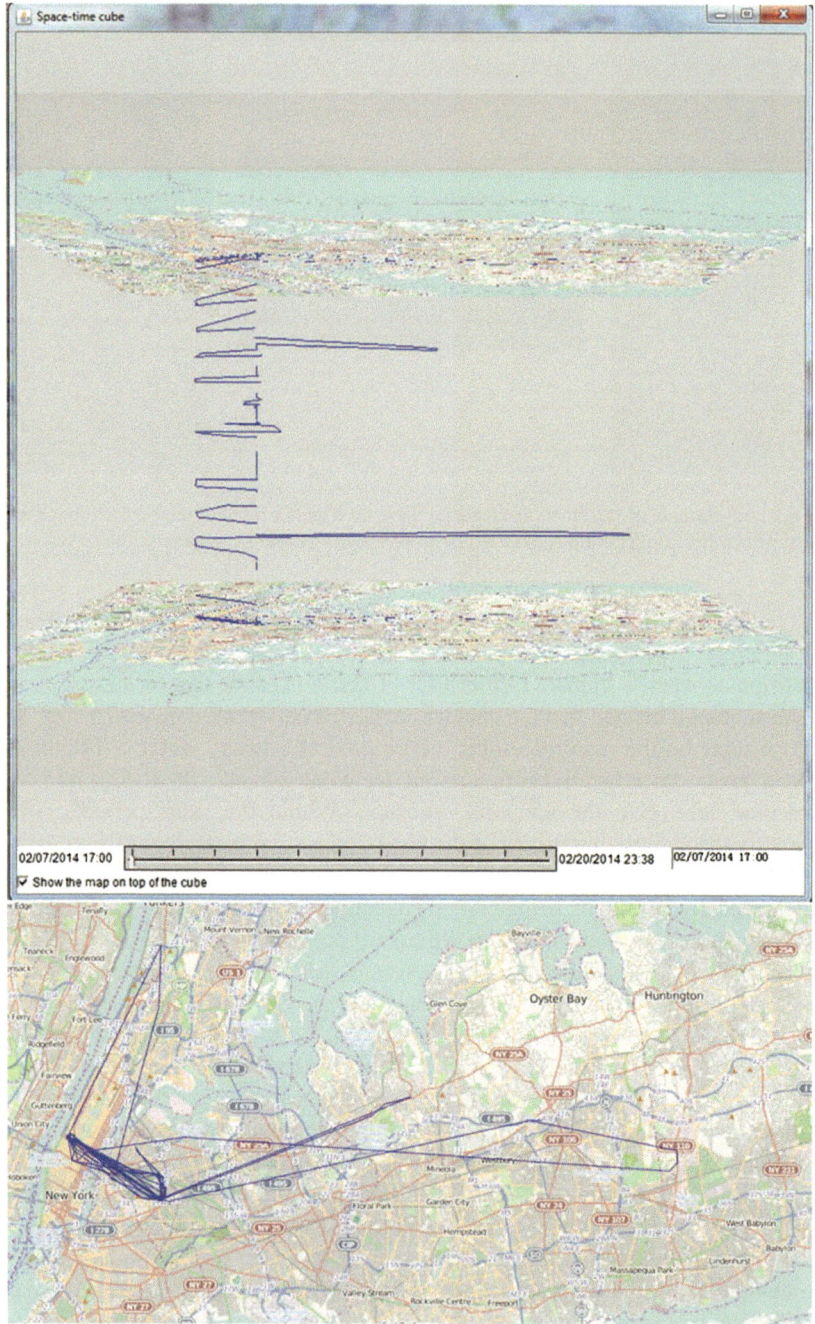

Fig. 6.2 Two space-time trajectories (left and right columns represent two users; the above windows show a 3D view and the bottom windows display a 2D geographical dimensions of the activities) left by human Twitter users (cited from Xu et al. 2015, courtesy of Taylor & Francis Ltd. http://www.tandfonline.com/)

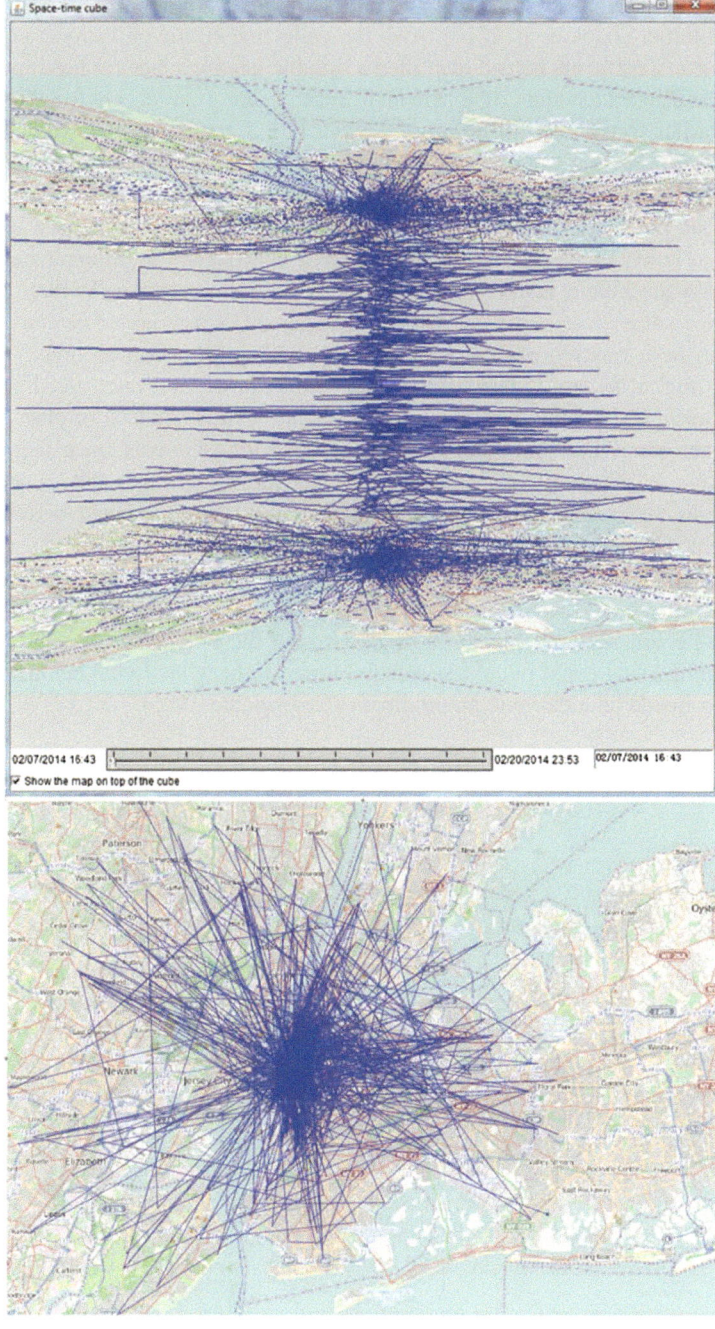

Fig. 6.2 (continued)

create a detailed transportation map of Haiti within days, if they were not connected by the Internet (Zook et al. 2010). Also the collective efforts of Tianya (a Chinese online forum) members helped identified a suitable landing place for helicopters at a remote village after the 2008 Sichuan Earthquake (Qu et al. 2009). In the humanitarian sector, the utilization of Twitter based approaches has demonstrated the great impacts that can be made by linking people to each no matter where they are (Meier 2011). Any messages sent to social media like Twitter would leave traces that potentially could be collected by other people. But it is a complex process to leverage the new data source. First, it is necessary to differentiate human users from machine users. Second, even though a human owns a Twitter account, the human subject can represent just a single person or a group of persons. Third, there can be a mixture of corporeal mobility with information mobility within a person's digital traces. Figure 6.3 presents place names excavated from a human Twitter users historical tweets. By referencing the textual contexts of place names, it becomes clear that some places were locations that the user once being there physically, while many other places were locations of the user's contacts on Twitter. In the second case a piece of information was exchanged between two geographic locations without the corporeal movement of the sender and the receiver. Thus, we need an identity structure that can consolidate the differences between human and machine subjects as well as between single-person account and multiple-person account. Inside the digital world the boundaries between human and machine, single and plural are greatly blurred to become digital beings capable of acting inside an electronic environment (Negroponte 1996).

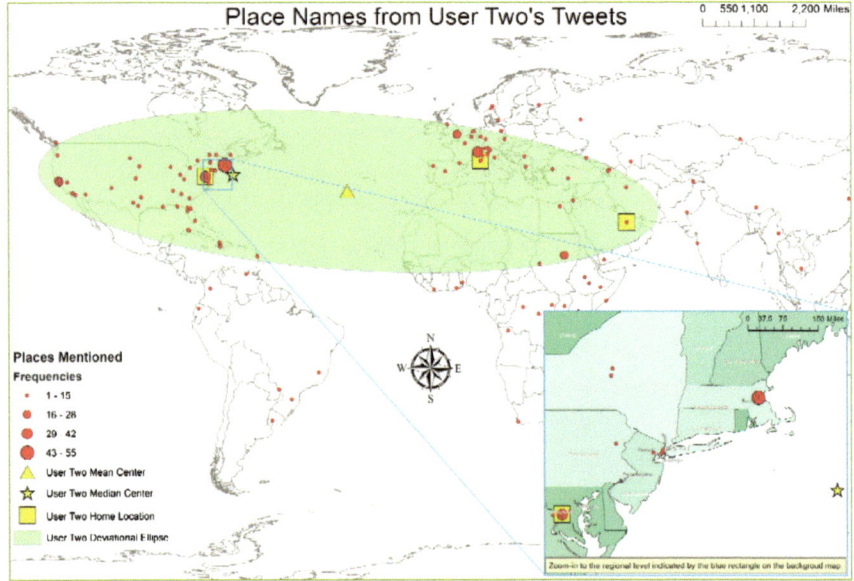

Fig. 6.3 Place names excavated from a human Twitter user's records (cited from Xu et al. 2013, courtesy of Taylor & Francis Ltd. http://www.tandfonline.com/)

A synergistic situation of bits and atoms creates a hybrid of part material and part immaterial, something called a cyborg, or a cybernetic organism. It was first a fiction creature like the robots with subjective experience in the movie *Blade Runner* (Nicholls 1984). Clynes and Kline (1995) proposed the term 'cyborg' for organisms augmented by external components for acquiring adaptive capabilities to environment. The concept also gets academic attention from several special study areas; Haraway (1991) applies the concept for feminism study, Mirowski (2002) synergizes the cyborg concept with economics, and in geography, Wilson (2009) leverages the cyborg theory for performing critical geographical analysis. With the dispersion and adoption of IoT through out people's daily lives, cyborg might reflect the future that we are moving towards (Gunkel 2000). Particularly, with cyberspace progressing from an ethereal existence that was previously floating above the physical space and was only attached to the physical world at fixed hubs toward a coexistence with the physical world via ubiquitous Internet connections, cyborg is not only a concept, but also one that deserves examination for its implications. Cyborg represents one realistic situation of human with augment of modern information and communication technologies (ICTs). People if not already cyborgs are living in a cyborg society (Gray 2000). For victims of recent Sony hack, the consequences are not just a horrible dream that vanished when waking up, but real threats to destroy their personal lives (The Economist 2014).

Here, the subjectivity of cyborg follows a traditional interpretation as an onto-logical hybridity, which consists of part organism and part machine (Wilson 2009). Firstly also majorly currently, the cyborg figure includes only the hybridity of human organism and ICTs; secondly, with scientific advancements in fields such as artificial intelligence that resulted in machines capable of recognizing cats,[3] the body of organisms might encompass a much versatile types of entities. Artificial Intelligence (AI) has gained profound progress recently as machines have demonstrated their capabilities to defeat the best players in games like Go. As AI increasingly expands its dominance in different facets of people's daily lives, the communicative processes and consequences become highly complex. The intension to put the phenomenon under the umbrella concept of cyborg is to highlight the intertwined effects of people's activities in the part-physical and part-virtual hybrid space. It is about this hybrid space that there is an astronomical amount of infor-mation. The democratization of web information production by the general users, for example, featured by the operation of Wikipedia, and the democratization of GIS first initiated by the availability of Google Earth and then boosted by the influential OpenStreetMap project incur an avalanche of collectable geographic data (Miller 2010; Dijck 2009; Goodchild 2007).

The aforementioned differences, e.g., between corporeal and information or single and plural, highlight some challenging aspects in studies that use the new types of data. As it is not always clear to what extent the machine and the organism collaborate in the endeavor, a unity, i.e., cyborg, is thus engaged. The hybrids are

---

[3]http://www.wired.com/2014/12/deep-learning-renormalization/.

capable of acting as one harmonious unit. The action can be automative, such as reports periodically sent by sensors, or can be communicative, such as human's or machine's engagement in exchanging messages. The implications of the new identity as cyborg could be enormous; "political-economic practices, the balance of class power" or the "cultural and social life" would be reoriented according to a cyborg's capacities to produce and consume one specific commodity, information. Extensive exploration of cyborgs is beyond the scope of this chapter. Here the author takes a practical and incremental approach to delineate the types of cyborg along three mechanical dimensions of the collectable raw data. New dimensions would be incorporated as suited to future demands. As illustrated by previous examples, the organism can be a single unit like individual people, or it can be a representation of an organization with multiple internal units. Then the movement of the cyborg can indicate location change of the material (i.e., corporeal) or the immaterial (i.e., information). Lastly subjective intension is considered stronger in the case when data is created manually compared to when automatically. Next a digital identity diagram presents the three dimensions of a cyborg's activities in a hybrid space (Fig. 6.4).

The cubic consists of three dimensions for defining the identity of a cyborg based on its form (corporeal vs. information), action (manual vs. auto) and structure (single vs. plural). Currently a majority of Twitter users can be recognized to be single human users that manually post tweets. These tweets can potentially reveal their producers' behaviors, but are different from traditional behavioral data from survey-based approaches. However, a person could be responsible for an organization account, in which information other than the corporeal movement is more important and the account represent a collection of internal corporeal units; thus in the Information/Single/Manual corner. A portion of Twitter accounts is shared by human members, thus the account identities should be put at the Information/Plural/Manual corner; and the mobility reflected in the movement of tweets is

Fig. 6.4 An identity cube for defining a cyborg's activities in a hybrid space

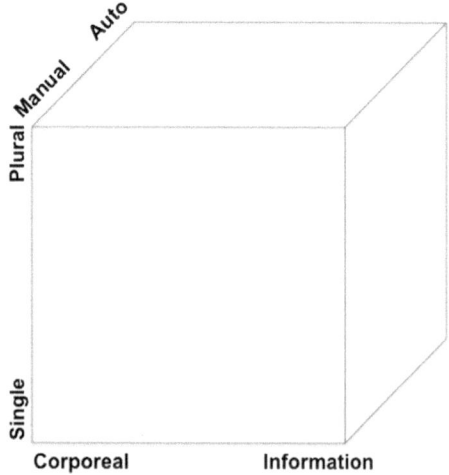

informational rather than corporeal. In the instance of virtual twitter users, some machines are configured to respond to external information requests therefore their behavior is more manual. Others are programmed to automatically post messages either regularly or when triggered by natural events such as an earthquake, thus their behavior is auto. Also to be potentially put at the Corporeal/Single/Auto corner are many wearable gadgets that monitor human health status or movements, which are capable of posting messages about their monitored subjects to Twitter.

6.5   A Cybernetic Linkage Between Micro and Macro Urban Cognitions

Another term for calling a cyborg is cybernetic organism. The behavioral nature of the entity is better denoted by this alternative name. An organism's action is dictated by the goal it seeks to realize. Cybernetics which is a science of purposeful systems was defined by Claude Shannon and Norbert Weiner's works almost at the same time (Shannon and Weaver 1949; Wiener 1948). Shannon and Weiner worked on different projects but their research on communication and control and the role of information in the process created the core of the field. While Shannon simply called the new field "information theory", Wiener created a mystical word "cybernetics".

Mechanically there is endless versatility of systems; cybernetics provides an ontological definition of systems about their essence. From cybernetics's perspective a system has defined goals to be achieved by utilizing its built-in functionalities. The system constantly adjusts its behavior until eventually hits the bullseye. During World War II, Weiner worked on automatic aiming and firing anti-aircraft guns, and from this work started to formulate the scientific field. The word 'cybernetics' was firstly created by Wiener and with a root in Greek word for 'steersman' (Wiener 1954). Weiner, different from Shannon, expanded the realm of cybernetics to beyond its engineering dimension, and suggested the theoretical framework applicable to examining phenomenon in fields such as biology, anthropology, and sociology.

Wiener saw the similarity between human and machinery in terms of utilizing the means of communication and control to adapt to external stimuli. An automatic anti-aircraft system senses the location of an airplane and repositions its aim accordingly without operators' involvement. Information is exchanged during the process. It is the mechanism of using information communication as to control behavior of a system that makes mechanical and biological systems comparable. The mechanism is a core concept in cybernetics, called feedback. Information technologies have been vehemently reshaping nature and society, and cybernetics provides a useful lens for examining the implications of the changes through the loop of communication, control, and feedback. Here the lens is fixed on cities and the dynamics of their dwellers. After the wikification of GIS, geographical information provides critical situational awareness in making smart decisions (Sui 2008). Poore and Chrisman (2006) proposed a social theory of geographic information based on cybernetic understanding of geographic information. Next the paradigm of cybernetics is applied for a critical analysis of geographic information in revealing human dynamics in cities. In general, techno-optimists view new technologies like big data computation, cloud computing, or AI as powerful force for making cities smart. It is hoped the inclusion of cybernetics would contribute to the understanding of the smart city technologies.

The formation of a communication, control and feedback loop is critical for a cybernetic system. A system issues a command for an intended behavior and assesses the effect by receiving feedback; both the command and the feedback are communicated as information. Weiner and many other cybernetics researchers demonstrated extensively individual humans follow cybernetic principles in daily activities. The same principles can also be applied to networked cities; for example, automation technologies are implemented to improve the efficiency of transportation networks or water networks of cities. The author of this chapter argues the becoming of cyborgs transforms the cognitive capacities of both the human technology users and the urban environment they dwell. Thus, a cybernetic linkage is created between the micro (individual cyborgs) worlds and the macro (cities) worlds. Next examples from Twitter users show possible linkages.

The social media platform of Twitter supports three very different communication modes simultaneously; the interactions among the three modes create a highly dynamic information sphere. Twitter users' interactions with others and reactions to broadcasted messages are captured by the platform and are materialized as explorable data. The digital traces left by interactivities of humans and machines create a complex digital world (Pickles 1995). Millions of Twitter users or cyborgs communicate with each other, receive information from connected machines, or broadcast to the wild digital world. Correspondingly three layers of communication exist simultaneously in the Twitter world. The more intimate conversations between two cyborgs are supported by using the "@" symbol with the information receiver's user name. Then information can be received by creating a follower network. Ultimately there is a broadcasting mechanism that can be used by anyone to send information to the entire Twitter world by using the "#" symbol with keywords (Bruns and Moe 2013).

The Twitter Company publishes sampled tweets through several APIs. Georeferenced tweets that have native location information are streamed through the geo API. By collecting tweets from this API, Zhang et al. (2015) draw a map of geographic distribution of mobile Twitter users in the region from the Washington DC to the New York City metropolitan areas. By analyzing traces left by these users, Jurdak et al. (2014) concludes that these data help reveal Twitter users' activities at the city level. Xu et al. (2015) identified several typical trajectory patterns that present activities in minute detail (see Fig. 6.2 for two examples). These trajectories present that people moving from one location to another in a city. The conglomeration of the apparent chaos at the individual level forms a harmonious rhythm of a city. Figure 6.5 shows a temporal distribution of total tweets by

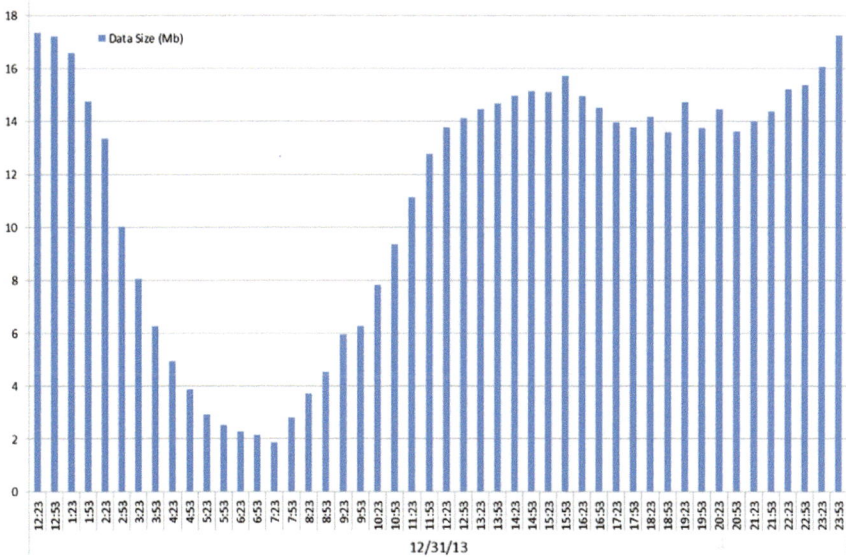

Fig. 6.5 Total number of tweets collected every 30 min during 24 h from New York City, NY

plotting the total number of tweets in terms of collected data size against every 30 min time frame for 24 h. The geographic area is New York City, NY.

For a city to become smart or more adaptive to events, it has to be aware of the city's status under various conditions. In cybernetics, a system needs not only to achieve its goal but also to maintain that goal under disturbance. If a city is looked as an organic cybernetic system and its dwellers become smart sensors, then the questions would be (1) whether the city has a stable status that can be quantified and (2) whether the effect of external disturbance can be detected by smart sensors and reflected as disturbance to the city's status.

When the total number of tweets is quantified and plotted in a 48-hour continuous fashion (Fig. 6.6), a rhythmic pattern emerges as the tweeting frequency rises up and wanes following almost identical temporal frame. However, at the turn of the last day of year 2013 there is a significant spike. It takes no effort to figure out that during those hours people were celebrating the beginning of year 2014. New Year's Eve is a special day and if it is considered as a disturbance event, apparently the collective behaviors of individual Twitter users (i.e., smart sensors) signify the moment by the total amount of greeting messages. Many other studies based on social media data, like from Flickr, also demonstrated the possibilities to use massive open data for measuring city's status. These measurements reveal different facets of cities; potentially controls could be imposed according to information sent by smart sensors to return cities to intended status.

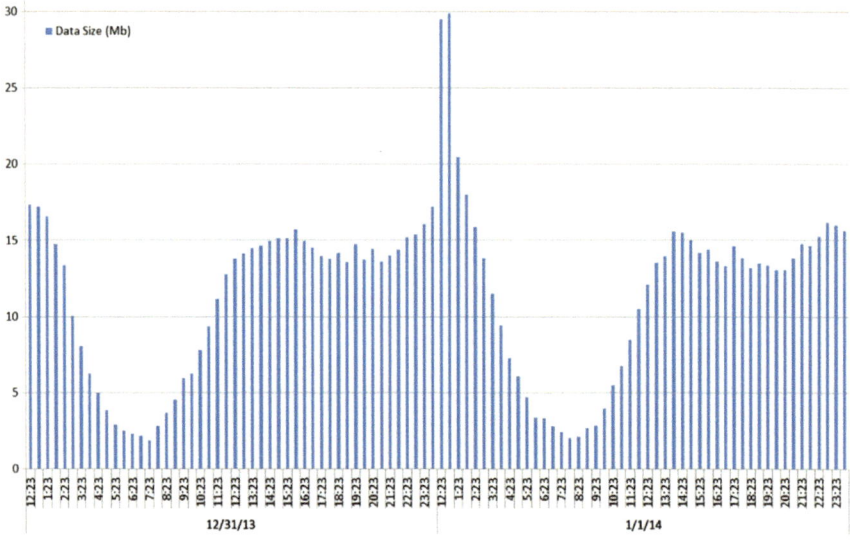

Fig. 6.6 Total number of tweets collected every 30 min during 48 h from New York City, NY

6.6   Project Cybersyn and the Cybernetic Possibility of Smart Cities

Looking cities as living organisms and applying cybernetic principles for optimizing cities' functionalities, or using a more recent phrase as smart cities, is after all an idea not new. There is historical experience that can be referred to for a glimpse of the future. One of the most ambitious cybernetic projects to date was Project Cybersyn, Chilean efforts to boost national economy through a cybernetic approach from 1971–1973 (Medina 2011). The project was never fully implemented and was terminated abruptly with the collapse of the Chilean government in a military coup. During the heyday of Cybersyn, a central control room with futuristic looking operation panels was built; a national wide telex network was laid out; economic data from different places of Chile were centralized to the control center; simulations were run to help experts in the command center for decision-making; and eventually control commands were dispatched to targeted places. Replacing the central control room with maybe a city's information center or a social media company's data center, the telex network with the Internet, economic data with far more versatile types of socioeconomic data, does the system sound familiar? Cybersyn project ended tragically because a technologically sound solution is not necessarily politically sound (ibid.). From a cybernetic perspective, a core teaching from Project Chilean is that a system's behavior is controlled by the built-in goals. From the beginning, Project Cybersyn followed Allende's government's determined goals for economic development (Salvador Allende was President of Chile during that period of time). Conflict interests were not resolved and led to the violence at the end. Undeniably drastic technology advancement has been made by human and it is likely not very far from delegating many city management activities to machinery with AI capability. The author argues that without socioeconomic structures for applying technology advancement for greater good of society, the cybernetic future of smart cities could be a slavery future for cyborgs. It is impossible to expand the discussion, but the author would like to call attention to several aspects of smart city technology at the end of the chapter.

Firstly, message is not equal to information. A cybernetic world is different from a Newtonian world and is governed by contingency. Both control and feedback in cybernetics are realized through communication. Communication channels have inherent noise. When command and feedback signals as information are transmitted through the channel, noise adheres to the information. Message received by a receptor of a cybernetic system is a conglomeration of information and noise. It is thus a critical challenge to recover information from the message. In other words, information is an organized structure. In the context of smart city, a possible implication is that big data is not equal to big amount of information. More important reading of this would come by applying the second order cybernetics, i.e., cybernetics of cybernetics. First order cybernetics is rather mechanical and does not fancy the subjective dimension of an information sender or receiver. Second order cybernetics brings the subjectivity into the cybernetic system, and makes the scientific principles more applicable to natures possessed by biological organism. Thus, information as an organized structure is a formulated intension. Without the existence of the second cybernetic loop, the intension might not be communicated. Consequently, the intension of the information sender may be misinterpreted by the receiver, or more seriously the inner cybernetic loop maybe manipulated intentionally.

To discuss the relevance of second order cybernetics to smart city technologies, let's look at a technology that is indispensible to enable the intelligence. As illustrated by Project Cybersyn, an economic decision was issued to a specific Chilean place to change the development at the targeted place, geographic location is one core technology to anchor the information layer to the physical environment of a city. Location is only the basic dimension of a place as defined by John Agnew (Agnew 1987). The other two dimensions are locale, the physical environment and the socioeconomic status of a place, and a sense of place, an entirely subjective perception of a place. Project Cybersyn already taught a lesson of a mechanical implementation of technologies to realize intelligence. The author would argue that smart city has to be not only an optimal technological structure but also a stable political condition. If smart city is interpreted as a second order cybernetic organism, the subjective factors can affect the entire technological realm of smart city. VGI is claimed to compensate authoritative geographic information with vernacular knowledge of a locale. Without technology support, the effect of VGI is filtered out by the inner cybernetic configuration.

Another interpretation of second order cybernetics is an observer-observed relationship. The inner cybernetic system is observed by an observer as the outer cybernetic system, and thence subjectivity become an influential element. The openness and accessibility of current information technology implies that an infinite number of first order cybernetic systems could be constructed out of the observer's intensions. In other words, out of people's sense of place, smart city can have very different meanings. Consequently, technology alone will not make city smart. There will be political struggles among different versions of a smart city. An ideal status out of the struggles would be a city that is robust to outer disturbance. But, the struggles may complete a city with a run-away condition or a perpetual oscillation status.

6.7   Summaries

Time geography can potentially facilitate a paradigm shift from population-based or place-based approaches to individual-based or people-based approaches and incorporate temporal dynamics into spatial analysis (Miller 2007; Peuquet 1994). The incorporation of time geography is thus important to the construction of smart cities. However, the technological realization of the conceptual design of a time geographic framework has been restricted by the technology capacity. With the enhancement in computing power, it becomes more practical to analyze human behaviors at the individual level that taking into consideration of personal variations (Armstrong 2004). Meanwhile, the increased Internet penetration into daily lives continues to transcend the real/virtual dichotomy to a digital-physical hybrid reality and cyborg is suggested to denote this unity (Coutrix and Nigay 2006; Fernback 2007). The physical space and cyberspace are co-constructed, and because of this hybridization, flow as a concept depicting the constant interactions between both spaces becomes important in understanding people's self-construction through place-making to concretize part of the space-time continuum as personal memories, experiences and many other forms (Leander and Mckim 2003). In order to accommodate human activities in hybrid spaces, the time geographic approach also has been modified (Shaw and Yu 2009; Yu and Shaw 2008). Recently, more studies have been focused on analyzing human dynamics by using Twitter data (e.g., Hasan et al. 2013; Cho et al. 2011). Compared to previous studies using mobile phone data, the using of Twitter data reveals more features about people's activities within and between cities (Jurdak et al. 2014). Potentially the integration of social media data would help reveal more dynamics happening in a city. But identity as a crucial part is not sufficiently addressed in most of these studies. There have been rich amount of studies on identities of social media users, such as Kietzmann et al. (2011) and Cho et al. (2011). Examining through the lens of cybernetics, the identity challenge indeed challenges the possibility of technology based smart cities.

Ellegard (1999, p. 168) cites Hägerstrand (1991) for arguing, "On a macro-level, human societies are, where they are located, formed by the continuous process of individuals performing activities in their everyday life." Traditionally to conduct quantitative studies on relations between societies and individual has been challenging because of the difficulties in data collection about human behaviors. The emergence of social media data, such as Twitter data, makes it possible to observe millions of people performing activities in their daily life. However, while the subject identity is generally unambiguous in traditional survey-based studies, the subject identity in Twitter based studies is complex. In order to accommodate the complexity, cyborg as an identity for entities that capable of acting in both a physical space and a cyberspace is proposed. Because of this dual capacity, when using the Twitter data for mobility study, attentions need to be paid to either the corporeal or the information dynamics is discussed. More importantly, methods need to be developed for categorizing the digital account into different groups. Although the task is challenging, there are elements that could be relied on for identification. For example, different spatial temporal patterns in data production or the construction style of the textual contents could be leveraged.

The conglomeration of digital traces can present an overall rhythmic pattern of urban life. The demonstration of an urban rhythm is valuable, as Lefebvre (1970) states that the analysis of rhythm would provide insight into the question of everyday life, the thing, the object, and life in the urban or rural environment. Meanwhile the individual activities that collectively form an urban rhythm are also collectable. Parkes and Thrift (1980) states that "in an urban context, the spatial distribution of facilities, their temporal availability (in terms of opening hours) and social institutions such as the working week conspire to create dynamic daily geographies of availability, or more accurately, presence and absence of opportunities." It is thence interesting to explore whether social media data in particular and open data generally could help shed light on the mobility patterns of individuals and their accessibility to opportunities. By digging into potential mechanisms that form the overall rhythmic patterns of urban dynamics out of millions of individual performing their daily activities at different geographic places of a city, it might help us understand the operations of cities which would lead to a future of smart cities. These insights can be valuable when urbanization is accelerating around the world. City is an organism with enormous complexity, if making city smart is a goal to be achieved, cybernetics offers a theoretical framework to critically examine the technological means for achieving such goal. The possibility of multiple versions of a smart city implies political struggles in order to dominate the future. Whether smart city is achievable would partly depend on if conflicts were consolidated in the development.

Acknowledgements The author would like to thank the anonymous reviews for their constructive comments and suggestions. Also, the author would like to express his appreciation to Drs. Shih-Lung Shaw and Daniel Sui for their great effort to organize this timely book.

References

Agnew, J. (1987). *Place and politics: The geographical mediation of state and society*. Allen and Unwin: Boston and London.

Ahas, R., Asas, N., Roose, A., Mark, Ū., & Silm, S. (2007). Evaluating passive mobile positioning data for tourism surveys: An Estonian case study. *Tourism Management, 29,* 469–486.

An, J., Cha, M., Gummadi, K., Crowcraft, J., and Quercia, D. (2012). Visualizing media bias through Twitter. In *Proceeding of ICWSM SocMedNews Workshop*. http://www.researchswinger.org/publications/an12visualizing.pdf. Last accessed on December 28, 2014.

Andrienko, G., Andrienko, N., Demsar, U., Dransch, D., Dykes, J., Fabrikant, S. I., et al. (2010). Space, time and visual analytics. *IJGIS, 24*(10), 1577–1600.

Armstrong, M. P. (2004). Geography and computational science. *Annals of AAG, 90*(1), 146–156.

Axhausen, K. W., Zimmermann, A., Schönfelder, S., Rindsfüser, G., & Haupt, T. (2002). Observing the rhythms of daily life: A six-week travel diary. *Transportation, 29,* 95–124.

Boyd, D., & Crawford, K. (2012). Critical questions for big data. *Information, Communication & Society, 15*(5), 662–679.

Bruns, A., & Moe, H. (2013). Structural layers of communication on Twitter. K. Weller, A. Bruns, J. Burgess, M. Mahrt, & C. Puschmann, C. (Eds.), *Twitter and Society*. Peter Lang, New York, 15–28.

Buckley, W. (1998). *Society: A complex adaptive system: Essays in social theory*. Amsterdam, The Netherlands: Gordon and Breach Publishers.

Buttimer, A. (1976). Grasping the dynamism of lifeworld. *Annals of the Association of American Geographers, 66*(2), 277–292.

Chechev, M., & Georgiev, P. (2012). A multi-view content-based user recommendation scheme for following users in Twitter. In *Proceedings of the 4th International Conference, SocInfo 2012* (December 5–7), Lausanne, Switzerland.

Cheng, Z., Caverlee, J., & Lee, K. (2010). You are where you tweet: A content-based approach to geo-locating Twitter users. In *Proceedings of CIMK'10* (October 26–30), Toronto, Ontario, Canada.

Clynes, M. E., & Kline, N. S. (1995). Cyborgs and space. *The cyborg handbook*, 29–34.

Cho, E., Myers, S. A., & Leskovec, J. (2011). Friendship and mobility: User movement in location-based social networks. In *Proceedings of the 17th ACM SIGKDD International Conference on Knowledge Discovery and Data Mining* (pp. 1082–1090).

Coutrix, C., & Nigay, L. (2006). Mixed reality: a model of mixed interaction. In *Proceedings of the working conference on Advanced Visual interface*, May 23–26, Venezia, Italy. New York, NY: ACM.

Cranshaw, J., Schwartz, R., Hong, J. I., & Sadeh, N. (2012). The livehood project: Utilizing social media to understand the dynamics of a city. In *Proceedings of the Sixth International AAAI Conference on Weblogs and Social Media*. http://www.aaai.org/ocs/index.php/ICWSM/ICWSM12/paper/download/4682/4967. Last accessed on November 27, 2014.

Dijck, J. (2009). Users like you? Theorizing agency in user-generated content. *Media, Culture and Society, 31*(1), 41–58.

Draijer, G., Kalfs, N., & Perdok, J. (2000). Global positioning system as a data collection method for travel research. *Transportation Research Record: Journal of the Transportation Research Board, 1719,* 147–153.

Ellegard, K. (1999). A time-geographical approach to the study of everyday life of individuals—A challenge of complexity. *GeoJournal, 48,* 167–175.

Elwood, S. (2009). Geographic information science: Emerging research on the social implications of the geospatial web. *Progress in Human Geography, 34*(3), 349–357.

Fernback, J. (2007). Beyond the diluted community concept: A symbolic interactionist perspective on online social relations. *New Media & Society, 9*(1), 49–69.

Garimella, V. R. K., Weber, I., & Cin, S. D. (2014). From 'I Love You Babe' to 'Leave Me Alone'—Romantic relationship breakups on Twitter. In *Proceedings of the 6th International Conference, SocInfo 2014* (November 11–13), Barcelona, Spain.

Goodchild, M. F. (2007). Citizens as sensors: The world of volunteered geography. *GeoJournal, 69,* 211–221.

González, M. C., Hidalgo, C. A., & Barabási, A.-L. (2008). Understanding individual human mobility patterns. *Nature* 453, 779–782.

Gray, C. H. (2000). *Cyborg citizen: Politics in the posthuman age.* New York: Routledge.

Greif, A. (1994). Cultural beliefs and the organization of society: A historical and theoretical reflection on collectivist and individualist societies. *Journal of Political Economy, 102*(5), 912–950.

Gunkel, D. (2000). We are Borg: Cyborgs and the subject of communication. *Communication Theory, 10*(3), 332–357.

Han, S. (2011). *Web 2.0.* London and New York: Routledge, Talyor & Francis Group.

Harvey, D. (1990). *The condition of postmodernity.* London: Blackwell.

Hägerstrand, T. (1991). *Tidens vidd och tingens ordning.* Carlestam: Texter sammanstallda ava B Sollbe och G.

Haraway, D. (1991). A cyborg manifesto. In *Simians, cyborgs, and women: The reinvention of nature.* Free Association Books, London.

Hasan, S., Zhan, X., & Ukkusuri, S. V. (2013). Understanding urban human activity and mobility patterns using large-scale location-based data from online social media. In *Proceedings of the 2nd ACM SIGKDD International Workshop on Urban Computing* (p. 6).

Hogan, B. (2010). The presentation of self in the age of social media: Distinguishing performances and exhibitions online. *Bulletin of Science Technology Society.* http://bst.sagepub.com/content/early/2010/10/31/0270467610385893.full.pdf+html. Last accessed on December 28, 2014.

Huang, Q., & Wong, D. W. (2016). Activity patterns, socioeconomic status and urban spatial structure: What can social media data tell us? *International Journal of Geographical Information Science, 30*(9), 1873–1898.

Isaacman, S., Becker, R., Cáceres, R., Kobourov, S., Martonosi, M., Rowland, J., et al. (2011). Identify important places in people's lives from cellular network data. In K. Lyons, J. Hightower, & E. M. Huang (Eds.), *Pervasive 2011*, LNCS 6696 (pp. 133–151). Verlag Berlin Heidelberg: Springer.

Janelle, D. G., Goochild, M. F., & Klinkenberg, B. (1988). Space-time diaries and travel characteristics for different levels of respondent aggregation. *Environment and Planning A, 20,* 891–906.

Jurdak, R., Zhao, K., Liu, J., AbouJaoude, M., Cameron, M., & Newth, D. (2014). *Understanding human mobility from Twitter.* arXiv prepring arXiv:1412.2154.

Jurgenson, N. (2012). When atoms meets bits: Social media, the mobile web and augmented revolution. *Future Internet, 4,* 83–91.

Kellerman, A. (1989). *Time space, and society: Geographical societal perspectives.* The Netherlands: Kluwer Academic Publishers.

Kendall, L. (1998). Meaning and identity in 'cyberspace': The performance of gender, class, and race online. *Symbolic Interaction, 21*(2), 129–153.

Kietzmann, J. H., Hermkens, K., McCarthy, I., & Silvestre, B. S. (2011). Social media? Get serious! Understanding the functional building blocks of social media. *Business Horizons, 54* (3), 241–251.

Kitchin, R. (2013). Big data and human geography: Opportunities, challenges and risks. *Dialogues in Human Geography, 3*(3), 262–267.

Kraak, M. (2003). The space-time cube revisited from a geovisualization perspective. In *Proceedings of the 21st International Cartographic Conference (ICC) 'Cartographic Renaissance'* (August 10–16), Durban, South Africa. http://hostmaster.icaci.org/files/documents/ICC_proceedings/ICC2003/Papers/255.pdf. Last accessed on November 26, 2014.

Kroeker, C. J. (1995). Individual, organizational, and societal empowerment: A study of the processes in a Nicaraguan agricultural cooperative. *American Journal of Community Psychology, 23*(5), 749–764.

Kwan, M.-P. (1999). Gender and individual access to urban opportunities: A study using space-time measures. *The Professional Geographer, 51*(2), 210–227.

Kwan, M.-P. (2004). GIS methods in time-geographic research: geocomputation and geovisualization of human activity patterns. *Geograflska Annaler: Series B, Human Geography, 86*(4), 267–280.

Kwan, M.-P. (2007). Mobile communications, social networks, and urban travel: Hypertext as a new metaphor for conceptualizing spatial interaction. *The Professional Geographer, 59*(4), 434–446.

Laurila, J. K., Gatica-Perez, D., Aad, I., Blom, J., Bornet, O., Do, T.-M.-T., et al. (2012). Presented at *Pervasive Computing*, Newcastle.

Leander, K. M., & Mckim, K. K. (2003). Tracing the everyday 'sittings' of adolescents on the Internet: A strategic adaptation of ethnography across online and offline spaces. *Education, Communication & Information, 3*(2), 211–240.

Lefebvre, H. (1970). *The urban revolution*. University of Minnesota Press.

McGee, E. M., & Maguire, G. Q., Jr. (2007). Becoming Borg to become immortal: Regulating brain implant technologies. *Cambridge Quarterly of Healthcare Ethics, 16*(3), 291–302.

McNeely, C. L., & Hahm, J. (2014). The big (Data) bang: Policy, prospects, and challenges. *Review of Policy Research, 31*(4), 304–310.

Medina, E. (2011). *Cybernetic revolutionaries: Technology and politics in Allende's Chile*. The MIT Press: Cambridge Massachusetts, London England.

Meier, P. (2011). New information technologies and their impact on the humanitarian sector. *International Review of the Red Cross, 93*(884), 1239–1263.

Miller, H. J. (1991). Modelling accessibility using space-time prism concepts within geographic information systems. *IJGIS, 5*(3), 287–301.

Miller, H. J. (1999). Measuring space-time accessibility benefits within transportation networks: basic theory and computational procedures. *Geographical Analysis, 31*(1), 1–26.

Miller, H. J. (2007). Place-based versus people-based geographic information science. *Geography Compass, 1*(3), 503–535.

Miller, H. J. (2010). The data avalanche is here. Shouldn't we be digging? *Journal of Regional Science, 50*(1), 181–201.

Miller, H. J., & Han, J. (2009). *Geographic data mining and knowledge discovery* (2nd ed.). Boca Raton, FL: CRC Press.

Minch, R. P. (2004). Privacy issues in location-aware mobile devices. In *Proceedings of the 37th Hawaii International Conference on System Sciences*. http://ieeexplore.ieee.org/stamp/stamp.jsp?tp=&arnumber=1265320. Last accessed December 28, 2014.

Mitchell, W. (2002). *Me++: The cyborg self and the networked city*. Cambridge, Massachusetts, London, England: The MIT Press.

Mirowski, P. (2002). *Machine dreams: Economics becomes a cyborg science* (p. 655). Cambridge: Cambridge University Press.

Negroponte, N. (1996). *Being digital*. Vintage.

Nicholls, J. G. (1984). Achievement motivation: Conceptions of ability, subjective experience, task choice, and performance. *Psychological Review, 91*(3), 328–346.

Noulas, A., Scellato, S., Mascolo, C., & Pontil, M. (2011). An empirical study of geographic user activity patterns in Foursquare. In *Proceedings of the Fifth International AAAI Conference on Weblogs and Social Media*. http://www.aaai.org/ocs/index.php/ICWSM/ICWSM11/paper/viewFile/2831/3241. Last accessed on November 27, 2014.

Oliver, M., Badland, H., Mavoa, S., Duncan, M. J., & Duncan, S. (2010). Combining GPS, GIS, and accelerometry: Methodological issues in the assessment of location and intensity of travel behaviors. *Journal of Physical Activity and Health, 7*, 102–108.

Parkes, D., & Thrift, N. J. (1980). *Times, spaces, and places: A chronogeographic perspective.* Wiley.

Peuquet, D. J. (1994). It's about time: A conceptual framework for the representation of temporal dynamics in geographic information systems. *Annals of AAG, 84*(3), 441–461.

Phithakkitnukoon, S., Horanont, T., Lorenzo, G. D., Shibasaki, R. & Ratti, C. (2010). Activity-aware map: Identifying human daily activity pattern using mobile phone data. A. A. Salah et al. (Eds.), HBU 2010, LNCS 6219 (pp. 14–25). Verlag Berline Heigelberg: Springer.

Pickles, J. (1995). Representations in an electronic age: Geography, GIS, and democracy. In *Ground Truth: The Social Implications of Geographic Information Systems* (pp. 1–30). New York: The Guilford Press.

Poore, B. S., & Chrisman, N. R. (2006). Order from noise: Toward a social theory of geographic information. *Annals of AAG, 96*(3), 508–523.

Qu, Y., Wu, P., & Wang, X. (2009). Online community response to major disaster: A study of Tianya Forum in the 2008 Sichuan Earthquake. In *Proceedings of the 42nd Hawaii International Conference on System Sciences*. http://ieeexplore.ieee.org/stamp/stamp.jsp?tp=&arnumber=4755366. Viewed on August 9, 2015.

Renshaw, J. R. (1976). An exploration of the dynamics of the overlapping worlds of work and family. *Family Process, 15,* 143–165.

Schaffers, H., Komninos, N., Pallot, M., Trousse, B., Nilsson, M., & Oliveira, A. (2011). Smart cities and the future internet: Towards cooperation frameworks for open innovation. *The future internet,* 431–446.

Shannon, C. E., & Weaver, W. (1949). *The mathematical theory of communication.* Urbana: University of Illinois Press.

Shaw, S.-L., & Wang, D. (2000). Handling disaggregate spatiotemporal travel data in GIS. *GeoInformatica, 4*(2), 161–178.

Shaw, S.-L., & Yu, H. (2009). A GIS-based time-geographic approach of studying individual activities and interactions in a hybrid physical-virtual space. *Journal of Transport Geography, 17,* 141–149.

Shaw, S.-L., Tsou, M.-H., & Ye, X. (2016). Editorial: Human dynamics in the mobile and big data era. *IJGIS, 30*(9), 1687–1693.

Soora, R. (2014). *Mobile technologies for activity-travel data collection and analysis.* Hershey, PA, USA: Information Science Reference.

Stopher, P. R., Bullock, P., & Horst, F. (2002). *Exploring the use of passive GPS devices to measure travel.* Institute of Transport Studies.

Sui, D. (2008). The wikification of GIS and its consequences: Or Angelina Jolie's new tattoo and the future of GIS. *CEUS, 32*(1), 1–5.

Sui, D., & Goodchild, M. (2011). The convergence of GIS and social media: Challenges for GIScience. *IJGIS, 25*(11), 1737–1748.

The Economist (2014). Horror movie: Hackers shine a harsh spotlight on Sony. http://www.economist.com/news/business/21636065-hackers-shine-a-harsh-spotlight-sony-horror-movie. Last accessed on December 24, 2014.

Tuan, Y.-F. (1977). *Space and place: The perspective of experience.* Minneapolis, MN: The University of Minnesota Press.

Urquhart, C., & Vaast, E. (2012). Building social media theory from case studies: A new frontier for IS research. In *Proceedings of the Thirty Third International Conference on Information Systems*, Orlando, FL.

Weber, J. (2003). Individual accessibility and distance from major employment centers: An examination using space-time measures. *Journal of Geographical Systems, 5,* 51–70.

Wiehe, S. E., Carroll, A. E., Liu, G. C., Haberkorn, K. L., Hoch, S. C., Wilson, J. S., et al. (2008). Using GPS-enabled cell phones to track the travel patterns of adolescents. *International Journal of Health Geographics, 7,* 22. http://www.ij-healthgeographics.com/content/pdf/1476-072X-7-22.pdf. Last accessed on November 26, 2014.

Wiener, N. (1948). *Cybernetics or control and communication in the snimal and the machine.* New York: Wiley.

Wiener, N. (1954). *The human use of human beings: Cybernetics and society.* Boston, MA: Da Capo Press.

Wilson, M. W. (2009). Cyborg geographies: towards hybrid epistemologies. *Gender, Place & Culture: A Journal of Feminist Geography, 16*(5), 499–516.

Wolf, J., Guensler, R., & Bachman, W. (2001). Elimination of the travel diary: Experiment to derive trip purpose from global positioning system travel data. *Transportation Research Record, 1768*, 125–134.

Wong, D. W. S., & Shaw, S.-L. (2011). Measuring segregation: an activity space approach. *Journal of Geographical Systems, 13*, 127–145.

Xu, C., Qin, H., & Yu, M. (2015). Visualizing spatiotemporal trajectories of mobile social media users using space-time cube. *CaGIS, 42*(1), 75–83.

Xu, C., Wong, W. D., & Yang, C. (2013). Evaluating the 'geographical awareness' of individuals: An exploratory analysis of Twitter data. *Cartography and Geographic Information Science, 40* (2), 103–115.

Ye, S. & Wu, S. F. (2010). Measuring message propagation and social influence on Twitter.com. In *Proceedings of the Second International Conference, SocInfo 2010* (October 27–29), Laxenburg, Austria,.

Yu, H., & Shaw, S.-L. (2008). Exploring potential human activities in physical and virtual spaces: A spatio-temporal GIS approach. *IJGIS, 22*(4), 409–430.

Zhang, H., Sun, Z., Liu, Z., Xu, C., & Wang L. (2015). Dart: A geographic information system on Hadoop. In *Proceedings of 8th IEEE international conference cloud computing* (June 27–July 2), New York, USA.

Zook, M., Graham, M., Shelton, T., & Gorman, S. (2010). Volunteered geographic information and crowdsourcing disaster relief: A case study of the Haitian Earthquake. *World Medical & Health Policy, 2*(2), 7–33.

Author Biography

Chen Xu (Ph.D., Texas A&M University), Assistant professor, University of Wyoming. His research interests focus on big geospatial data analytics, geospatial cyberinfrastructure, volunteered geographic information, and geographic information science. Xu used big data computational technologies for processing large amount of digital traces left by social media users like Twitter users for mapping individuals' mobility patterns and analyzing spatial temporal evolvements of geographic events. He also worked on exploring the theoretical aspects of studying the dynamics between people and technologies so as to make place out of the synthesis of physical space and cyberspace.

Chapter 7
A Location-Based Client-Server Framework for Assessing Personal Exposure to the Transmission Risks of Contagious Diseases

Tzai-Hung Wen, Ching-Shun Hsu, Chih-Hong Sun, Joe-Air Jiang and Jehn-Yih Juang

7.1 Introduction

Due to the complex interactions between human behaviors and the environment, it is important to quantify the association between environmental exposure and human health. Many studies have focused on identifying the spatial distribution and risk factors of vulnerable populations by analyzing the exposure to a risk. These results are often used to develop intervention programs (Weis et al. 2005). Previous studies have shown that understanding personal mobility and routine behaviors are key factors to assess individual environmental risk exposure (Gerharz et al. 2009; Kwan 2009; Perchoux et al. 2013). These studies have indicated that different behaviors by an individual may result in different levels of risk exposure. Therefore, different personal behaviors should be incorporated into a framework used to assess risk exposure (Kwan 2009; Setton et al. 2011). Recently, the rapid improvements to information and communications technology (ICT), such as biological sensors, geographic information systems (GIS) and global positioning systems (GPS), have made it possible to quantify individual behaviors more comprehensively. Recent studies have used ICT technology to quantify the effects of individual behaviors on the risk of exposure to air pollution and volatile organic compounds (VOCs) (Fang and Lu 2012; Leyk et al. 2009; Wang et al. 2009). These studies have improved the differentiation of the types of individual mobility and routine behaviors in time and space and have provided a detailed understanding of the risk of exposure to environmental pollution (Gerharz et al. 2009; Dons et al. 2012; Steinle et al. 2015). These findings provide an important reference for the designations of health policies and environmental regulations (Gerharz et al. 2009; Leyk et al. 2009; Steinle et al. 2011).

T.-H. Wen (✉) · C.-S. Hsu · C.-H. Sun · J.-A. Jiang · J.-Y. Juang
National Taiwan University, Taipei, Taiwan
e-mail: wenthung@ntu.edu.tw

© Springer International Publishing AG, part of Springer Nature 2018
S.-L. Shaw and D. Sui (eds.), *Human Dynamics Research in Smart and Connected Communities*, Human Dynamics in Smart Cities,
https://doi.org/10.1007/978-3-319-73247-3_7

Most ICT-based studies focus on collecting behavioral data on individuals to identify the association between individual behaviors and environmental exposure. However, it is still unclear how to use ICT-based data to assess the exposure of an individual to contagious infection risks, such as influenza or the measles (Weis et al. 2005; Perchoux et al. 2013). Contagious diseases are often transmitted by close person-to-person contact. ICT-based data could capture mobility trajectories and interpersonal interactions, which are important characteristics for understanding the spread of infectious diseases (Stoddard et al. 2009). The mechanism of contagious disease transmission involves complex social interactions and is therefore different from the risk of exposure to environmental pollution. For example, acute contagious diseases, such as H1N1 or SARS, have shorter latent periods and promptly cause serious clinical symptoms. Moreover, the risk of exposure to air pollution and volatile organic compounds (VOCs) is mainly caused by the exposure to the environment and is not related to person-to-person contact. Complex human contacts cause an exponential increase in the transmission risk during the initial period of transmission of contagious diseases such as influenza A (H1N1) (De Silva et al. 2009; Lessler et al. 2009). Therefore, the risk of exposure to contagious diseases may be proportional to the frequency and duration of human contact or visits (Rothman et al. 2008). In other words, human mobility behavior is a key factor for measuring the risk of exposure to contagious diseases. Different human contact structures also influence the patterns of disease transmission (Funk et al. 2009; Meloni et al. 2011; Wang et al. 2012; Wu et al. 2012). Therefore, incorporating human mobility and person-to-person contact are necessary to measure personalized risks of exposure to contagious diseases.

Previous studies on exposure risk assessment have focused on quantifying the relationships between personal contacts and exposure dose (Dons et al. 2011, 2012) and modeling the spatial distributions of exposure risk (Gerharz et al. 2009; Leyk et al. 2009; Steinle et al. 2011; Wang et al. 2009; Liu et al. 2013). These studies have been based on the decisions of individual behaviors and could be beneficial for reducing personal exposure risk (Gerharz et al. 2009). However, other recent studies have used smartphones as a mobile health platform to improve the timely delivery of medical advice and health communication between health care workers and patients (Martin et al. 2016; Stanton et al. 2016). Moreover, recent studies have also focused on the interactions between human behaviors and disease transmission (Mao 2014; Zhao and Wu 2014). People may change their behaviors to avoid infection if they realize the increased risk of being infected during an epidemic period. Meanwhile, changes to individual behaviors may also provide feedback on the dynamics of transmission. For example, Mao (2014) established an agent-based model to integrate information, disease diffusion and behavioral changes and simulate these three diffusion aspects in metropolitan areas. These studies have concentrated on measuring the impacts of behavioral changes on disease diffusion, although the mechanisms of these interactions remain unclear (Funk et al. 2009; Wu et al. 2012; Kiss et al. 2013; Perra and Vespignani 2013). Perra and Vespignani (2013) used network approaches to create a behavior-disease meta-population model to understand the mechanisms of different societal reactions that trigger changes in the mobility behavior of individuals. Most of these studies have used computer modeling approaches to simulate the impacts of behavioral changes of individuals on disease transmission. However, how to broadcast epidemic information to individuals effectively and monitor their behavioral changes promptly have not been fully investigated or implemented. Applications (apps) on smartphones are now an appropriate low-cost platform that can be used to collect personal behavior data and transfer digital information between individuals (English 2016).

The client-server framework integrates with the transmission dynamics of the disease simulation models and can reflect transmission risks due to behavioral changes in a timely manner. Apps on smartphones are an appropriate platform for individuals to effectively collect information on changes in personal mobility behaviors. Through a location-based framework, individuals can also receive timely risk alerts to assess their movement patterns and reduce their risks of exposure to infection. Therefore, the purpose of this study is to establish a location-based client-server architecture to assess the exposure of an individual to the risk of contagious disease transmission. The specific aim is to integrate transmission dynamics with mobility behaviors to develop a personalized exposure assessment framework that will broadcast the individual risks of exposure to infection in a timely manner. A hypothetical scenario of an influenza epidemic at a university campus was used as a pilot study to demonstrate the feasibility of a location-based client-server framework.

7.2 Methods

7.2.1 The System Architecture for Personalized Exposure Assessment

A client-server architecture is adopted for the framework for personalized exposure assessment (Fig. 7.1) and consists of two major components. The first component is the is client-side smartphone-based exposure assessment module. We developed an Android application to collect course enrollment records and real-time location data to display personalized exposure scores. The second component consists of the server-side database and the simulation model. The simulation model calculates the personalized exposure score based on the course enrollment records or the real-time GPS logs of individual mobility data from the client-side of the Android application.

We used the main campus of National Taiwan University (NTU) in a pilot study to demonstrate the feasibility of the framework. Attending classes results in long durations of face-to-face contact in an indoor environment and could increase the risk of contagious disease transmission. Therefore, this study focused on analyzing the spatial structures of class attendance behaviors. To represent the class attendance behavior, we used class enrollment data from a major university in Taiwan. The records of the students enrolled in courses were used to build the origin-destination (OD) matrix used to capture the mobility of students around the

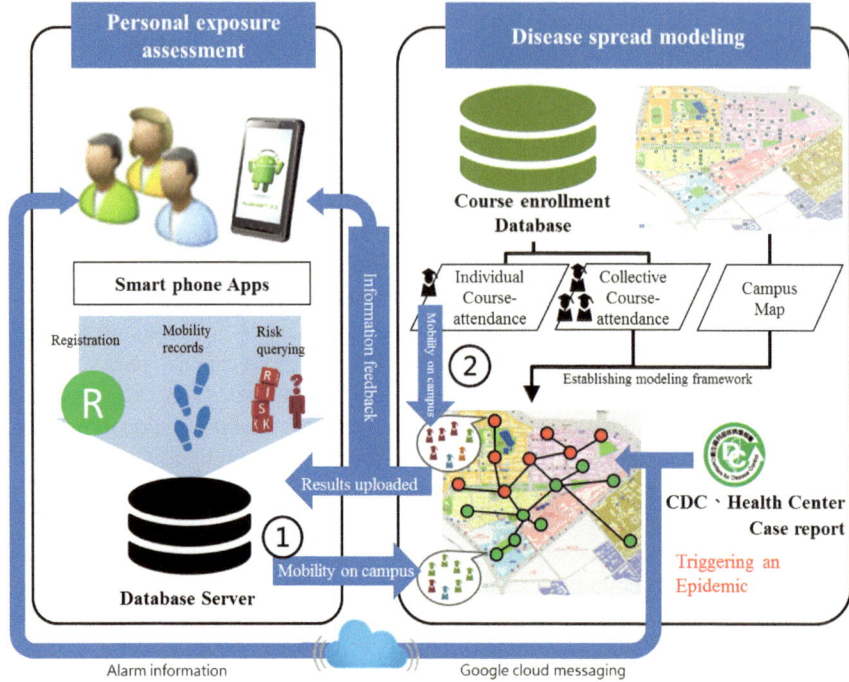

Fig. 7.1 The client-server framework for assessing the personalized risk of exposure to contagious diseases

campus. We then modeled the spatial interaction relationships among classroom buildings. In addition, we also developed a GPS tracking module to record the temporal and spatial mobility history of individuals in case they changed their course schedule or were absent from classes. The exposure risk was measured based on the parameters of disease progression and the contact frequency from the individual mobility matrix. The flowchart in Fig. 7.1 illustrates that the server-side epidemic simulation is triggered when a student is reported by the routine communicable disease reporting procedures of the Taiwan Centers for Disease Control (Taiwan-CDC) or local health centers. Each student who installed the client-side risk assessment module on their smartphone receives a personalized exposure score when an epidemic outbreak occurs on the NTU campus.

7.2.2 Server-Side Epidemic Dynamics Simulation Model

The epidemic simulation consists of two components: (a) the building space-time mobility network; and (b) the simulation of the progression of an epidemic. The detailed procedure is described below.

7.2.2.1 Space-Time Mobility Network

The class enrollment data included several tables: a list of students and departments, a list of departments and buildings and student course enrollment records. Because we focused on the disease transmission within department buildings, outdoor courses were excluded from the dataset. Moreover, classes held on other campuses were also excluded. Thus, all indoor lecture/seminar courses that occurred at the main campus were included in our dataset. Our dataset included 24,975 students, 3214 courses, 190,469 registration records, 6059 classes, the locations of 75 department buildings, and the relevant information for these departments. We summarized the data into several tables in a database, including the total number of students who took classes in each building, the number of students moving between each department building, and the geographical distance between each building. To capture the potential contact frequency among students in different departments each day, we summarized the different locations of classroom buildings for each student from Monday to Friday to capture the mobility heterogeneity within one week. The study constructed geo-referenced location networks based on the students' class locations and the degree of mobility among department buildings in terms of the OD matrix. Furthermore, to capture the mobility heterogeneity within each day, the matrix was further decomposed into different mobility networks in two time-slices: morning (before 13:00) and afternoon (after 13:00). Finally, we constructed a mobility network with 10 time-slices (2 time-slices/day × 5 days) for one week for each student, which captures the variations within one day and between days in a week (Fig. 7.2).

7.2.2.2 Simulating the Progression of an Epidemic

The four-state Susceptible-Latent-Infective-Removed (SLIR) mathematical model, which integrates space-time mobility networks, was built to simulate the progression of an epidemic and determine the exposure score for each classroom building. We proposed a network-based meta-population model to simulate the space-time diffusion of an epidemic (Hess et al. 2001). The model is composed of various subpopulations as nodes in a network, and the links between the nodes can represent the mobility of the individuals moving among these nodes (Meloni et al. 2011; Perra and Vespignani 2013; Tsai et al. 2010). In our pilot study at the campus, a subpopulation is defined as a classroom building, and a link is defined as the degree of mobility between classroom buildings. The epidemic status of an individual is initially set at Susceptible (vulnerable to infection but not yet infected), followed by Latent (infected but unable to infect others), Infective (capable of infecting other individuals), and Removed (i.e., recovered, deceased, or otherwise not posing any further threat).

The number of pathogens that Susceptible-to-Latent hosts carry are insufficient to actively transmit to other Susceptible hosts. These numbers eventually reach levels where hosts become Infectious, begin to infect other Susceptible hosts, and

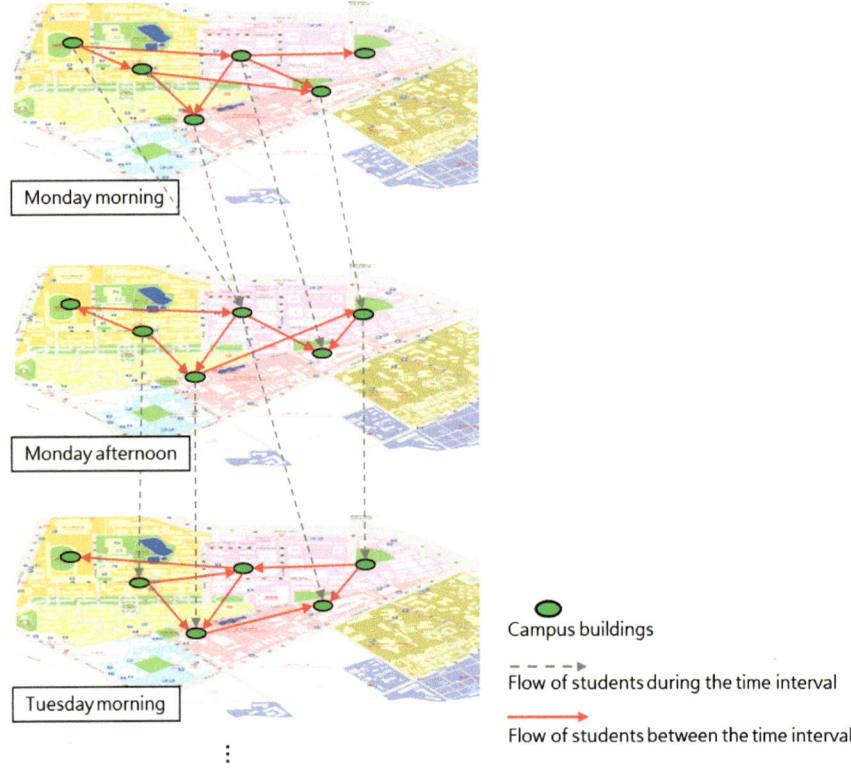

Fig. 7.2 A conceptual diagram of the space-time individual mobility network

eventually move toward a Removed status. The dynamics of these four epidemic states of classroom building i, $S_i(t)$, $L_i(t)$, $I_i(t)$ and $R_i(t)$, over time are expressed as Eqs. 7.1–7.4, with the following descriptions.

$$\frac{dS_i(t)}{dt} = \frac{-\beta S_i(t) I_i(t)}{N_i} - \sum_{j \neq i} \beta S_i(t) \frac{W(j,i)(t)}{\sum_k W(j,k)(t)} \frac{I_j(t)}{N_j} \tag{7.1}$$

$$\frac{dL_i(t)}{dt} = \frac{\beta S_i(t) I_i(t)}{N_i} + \sum_{j \neq i} \beta S_i(t) \frac{W(j,i)(t)}{\sum_k W(j,k)(t)} \frac{I_j(t)}{N_j} - \theta L_i(t) \tag{7.2}$$

$$\frac{dI_i(t)}{dt} = \theta L_i(t) - \gamma I_i(t) \tag{7.3}$$

$$\frac{dR_i(t)}{dt} = \gamma I_i(t) \tag{7.4}$$

N_i is the total population of classroom building i. The transmission rate β is a constant representing the transmissibility of a communicable disease. Susceptible individuals become Infected and acquire a Latent status (no clinical symptoms and cannot infect others). The latent rate θ is a constant used to determine the latent period from Latent to Infective status (fever onset and begin to infect others). The removed rate γ is a constant used to determine speed of transformation from Infected to Recovered. The outbreak of influenza A/H1N1 was simulated as the hypothetical scenario in this study. The parameters of the epidemic progression were set as $\beta = 0.585$, $\theta = 0.32$, $\gamma = 0.09$ (Samanlioglu et al. 2009). We simulated the transmission dynamics of the influenza A/H1N1 over 120 days after the first infected student was identified.

To capture the frequency of contact among students in different classroom buildings, the proposed SLIR model was used to reflect the effect of space-time mobility (Eq. 7.1). For all i and j locations in a mobility network W(j, i), we propose the concept of the geodemographic weight shown as $\frac{W(j,i)(t)}{\sum_k W(j,k)(t)} \frac{I_j(t)}{N_j}$ to measure the frequency of contact among students in different classroom buildings. W(j, i) indicates the population flow from the locations of classroom buildings j to i at time-slice t. The geodemographic weight refers to the proportion of the infected students in classroom building j who move to building i. $\beta S_i(t)$ indicates the number of infected cases in the classroom building i that results from one infectious person at time-slice t. Therefore, the result of the multiplication in Eq. 7.1 represents the transmission risk of classroom building i due to the mobility of students around the campus. Moreover, the force of infection (λ) is defined as the rate at which susceptible individuals become infectious (Balcan et al. 2010; Sattenspiel 2009). Since the indicator captures the intensity of the disease transmission, it is used as the score of the exposure to the infection risk in classroom building i at time-slice t (Eq. 7.5).

$$\lambda_i(t) = \frac{\beta \times I_i(t)}{N_i} \tag{7.5}$$

7.2.3   Client-Side Personalized Risk Assessment Module

We used smartphones as the platforms for the client-side risk assessment. The platforms include (a) service registration; (b) the records of individual mobility, and (c) the personalized exposure score.

7.2.3.1   Service Registration

We developed the service registration module to provide personalized service. Each user account and their smartphone identifier, International Mobile Equipment

Identity (IMEI), are created in a back-end relational database with the personal account. Each user account can provide personalized services based on course schedules, mobility behaviors, and exposure scores.

7.2.3.2 The Records of Individual Mobility

The smartphone app provides two modes for recording individual mobility: self-reporting and GPS logs. Users can self-report their locations and the time of visit using a map-based user interface for data input. Otherwise, users can use their GPS receivers in the smartphones to record their locations. Both self-reporting data and GPS logs transfer to the back-end database. The transferred information includes the locations and time of the visit, the visit duration, and the cell phone identifer IMEI code. Since the IMEI code is unique, we only use the code to reflect the different behaviors among individuals rather than request any personally identifiable information, such as name or gender.

7.2.3.3 The Score of Personalized Exposure to the Infection Risk

The exposure score in building i in one time-slice t, $\lambda_i(t)$, can be measured as the force of infection (Eq. 7.5). Based on the records of mobility behavior of one student k, we can create the list of buildings visited $B_k = \{i_1, i_2, i_3 \ldots, i_n\}$, visit time $T_k = \{t_1, t_2, t_3 \ldots, t_n\}$ and visit duration (hours) $D_k = \{d_1, d_2, d_3 \ldots, d_n\}$. Then, we can determine the exposure in each building $R_k = \{\lambda_1, \lambda_2, \lambda_3 \ldots, \lambda_n\}$, and then weight the exposure by visit duration $R_k \times \frac{D_k}{24}$. Figure 7.3 illustrates one student that traveled among the buildings with different exposure scores. We can use the trajectories of mobility of an individual to determine his/her exposure to the infection risk. If student k visited n buildings within one day, the probability of not having any exposure to the risk could be calculated as $\prod_{i=1}^{n} \left(1 - R_{k,i} \times \frac{D_{k,i}}{24}\right)$. Therefore, the exposure score of student k for the day is expressed by the following equation (Eq. 7.6).

$$E_k = 1 - \prod_{i=1}^{n} \left(1 - R_{k,i} \times \frac{D_{k,i}}{24}\right) \qquad (7.6)$$

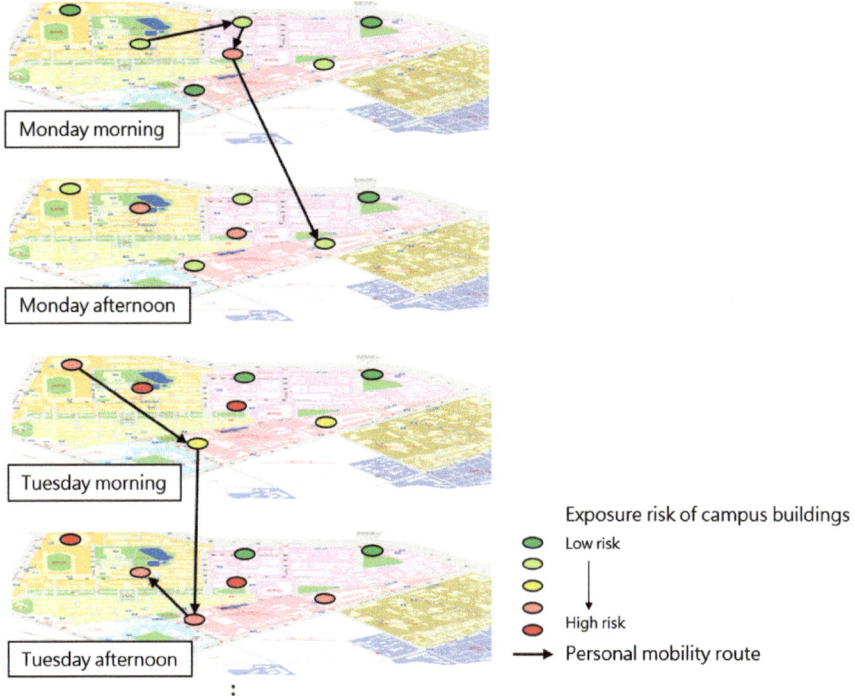

Fig. 7.3 An example of the tracking of the mobility of an individual to measure the personalized exposure to the infection risk

7.3 Results and Discussion

The user case diagram is illustrated to demonstrate the interactions between users and the system architecture in Fig. 7.4. A User, such as a student, registers their account in the smartphone app, and the app then collects their mobility behaviors and transfers this information to back-end database via WIFI or telecommunication. The users ultimately receive the score of their exposure to the risk based on their mobility behaviors when there is an epidemic outbreak. The competent authorities, such as personnel at the health center, receive the confirmed cases from the disease reporting system and trigger the transmission dynamics. The estimated transmission risk of each building can be uploaded and monitored by the user's smartphones. We show the following results of the different scenarios based on user-side and competent authority-side applications.

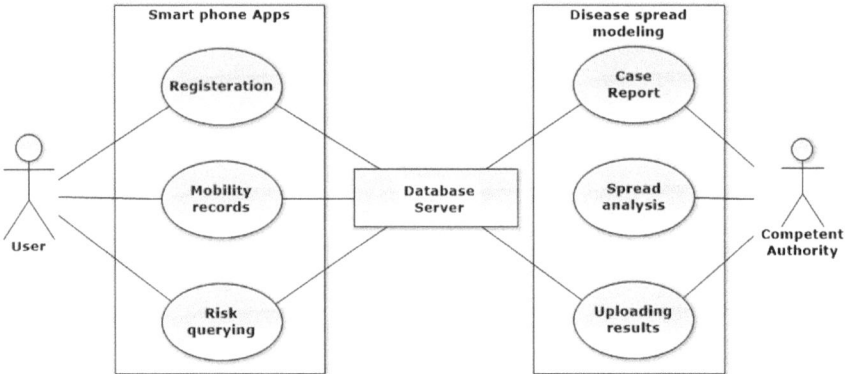

Fig. 7.4 A user case diagram demonstrating the interactions between the users and the system architecture

7.3.1 User-Side Application

The personalized exposure to the risk is measured by either course schedule or mobility records. Figure 7.5 show the results of different modes for visualizing the spatial distribution of the exposure scores. Figure 7.5a illustrates the personalized exposure to the risk based on an individual's course schedule. The app also reports the predicted risk over the next two days if the individual chooses to attend these courses. Figure 7.5b illustrates the spatial distribution of the transmission risk in each classroom building. Based on the risk map, an individual can choose to visit the classroom buildings with lower transmission risks or avoid traveling near the buildings with high risks.

In addition, the app also reports the risk based on the mobility behaviors from the self-reporting or GIS logs in Fig. 7.5c. If a student appears with symptoms of acute contagious diseases during the epidemic period, such as a cough or sneeze, they can check the mobility trajectories around the campus over the past few days. The app then reports the estimated exposure to the risk based on the mobility data. The student could realize the possibility of getting infected on the campus. A higher exposure score may represent a higher association between the mobility patterns of the student and the risk of disease transmission.

7.3.2 Competent Authority-Side Applications

The results of the simulated transmission dynamics could also provide a competent authority with the overall epidemic progression and the spatial and temporal distribution of the transmission risk. This information could be importation for policy interventions. For example, Table 7.1 shows the NTU guidelines for epidemic

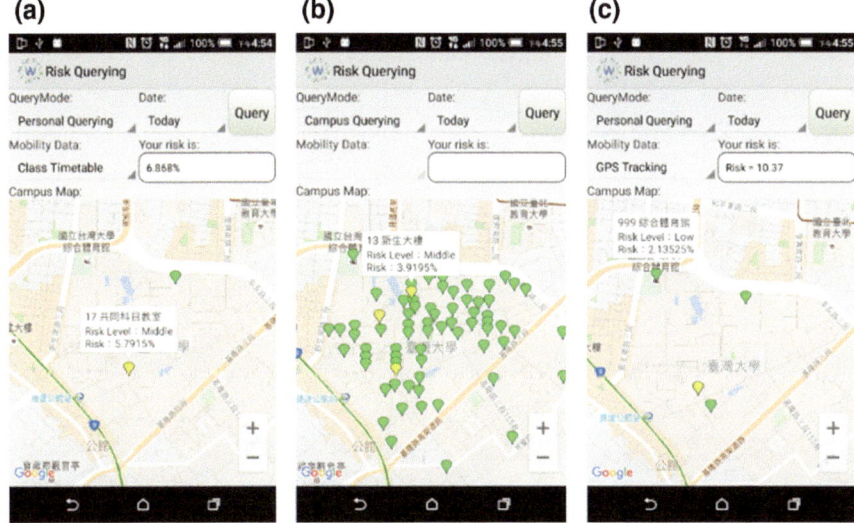

Fig. 7.5 Example scenarios from the user-side applications

Table 7.1 NTU Guidelines for epidemic control measures

Epidemic level		Control measures
Phase 1	No case in the country	Self-health management
Phase 2	Cases have occurred in the country but no case on campus	1. fever screening 2. disinfection of instruments 3. avoidance of large-scale activities
Phase 3	Cases have occurred sporadically on the campus	1. health authorities enter and are stationed at the university 2. arrange quarantine locations
Phase 4	Clusters of cases on the campus	1. suspend classes or close the school 2. forbid large-scale activities

control measures based on different levels of epidemic alerts during the 2009 influenza/H1N1 pandemic. Phases 1 to 4 describe the different levels of an epidemic and determine which control measures to implement, such as disinfection of instruments and hands, quarantines or school closure. The temporal and spatial estimated risk provides valuable clues to implement the appropriate measures. Figure 7.6 shows the results of the scenarios from the authority-side applications. Figure 7.6a shows the four-stage population dynamics, and the user can determine the predicted scale of the epidemic peak and the estimated time to the peak. Figure 7.6b shows the spatial-temporal distribution of the transmission risks in each classroom building. Based on the map, the campus could be further partitioned into zones with different risk levels, and the spatial clustering of the high-risk areas can be identified. The classroom buildings in these focused areas could be considered

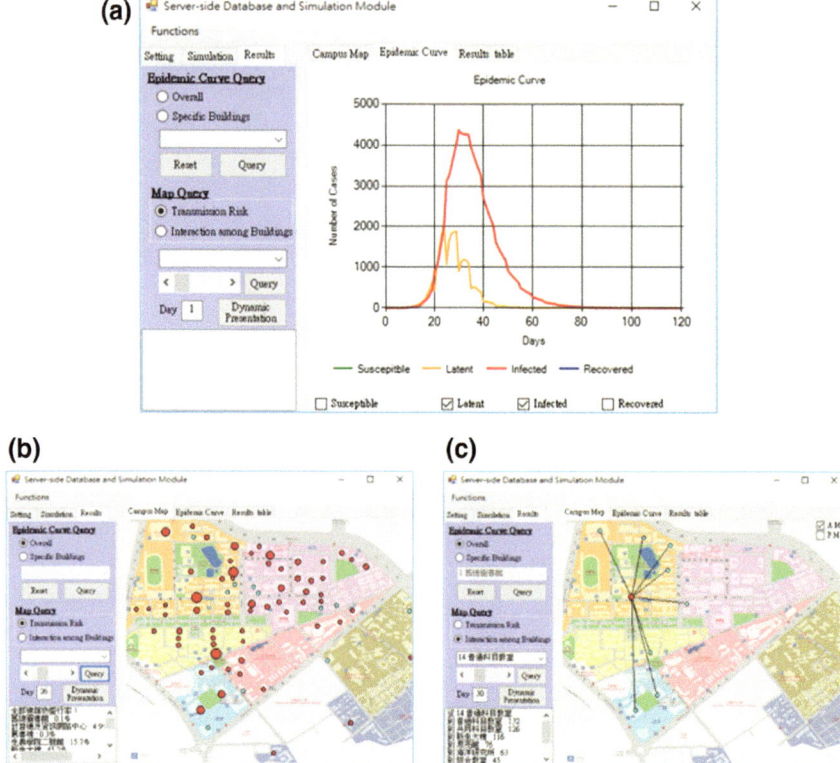

Fig. 7.6 Example scenarios from the authority-side of the application

for high-intensity disinfection or suspended classes. Figure 7.6c illustrates the intensity of the interaction among buildings. The nodes and links in the map show the high degree of mobility of students among these buildings. In other words, if clustered infected cases occur in the red node, the blue nodes have the highest priority for high-intensity disinfection to block the transmission chain.

7.4 Conclusion

Based on individual mobility behaviors, this study proposed a location-based client-server framework to measure personalized exposure to the transmission risks of contagious diseases at a university. Classes are the most common indoor activity and result in face-to-face contact and increased transmission risk of acute contagious diseases. We used a course enrollment database to generate a spatial and temporal mobility network to reflect the heterogeneity of the contact frequency of students among the classroom buildings. In addition, our framework allowed for individuals to self-report their locations or use GPS logs to capture more accurate information on mobility behaviors. Therefore, based on the epidemic simulation and the individual mobility network, each student who uses their smartphone as a personalized platform can understand the progression of the epidemic and make better behavior decisions, such as wearing a face mask or reducing their contact frequency, based on the personalized exposure scores from the server-side computation. In sum, the framework presents complex interactions among the reception of personal risk, behavioral changes, and epidemic progression. More scenarios can be implemented in a future study to quantify the effects of risk reception or behavioral changes on the progression of an epidemic.

Acknowledgements The research was supported by the grants of the Ministry of Science and Technology in Taiwan (MOST 103-2627-M-002-006, MOST 104-2627-M-002-020-). The authors also acknowledge the financial support provided by Infectious Diseases Research and Education Center, Ministry of Health and Welfare (MOHW) and National Taiwan University (NTU). The funders had no role in the study design, data collection and analysis or in the preparation of the manuscript.

References

Balcan, D., Goncalves, B., Hu, H., Ramasco, J. J., Colizza, V., & Vespignani, A. (2010). Modeling the spatial spread of infectious disease: the global epidemic and mobility computational model. *Journal of Computation Science, 1*(3), 132–145.

De Silva, U. C., Warachit, J., Waicharoen, S., & Chittaganpitch, M. (2009). A preliminary analysis of the epidemiology of influenza A(H1N1) v virus infection in Thailand from early outbreak data, June–July. *Eurosurveillance, 14*(31), 19292.

Dons, E., Panis, L. I., Van Poppel, M., Theunis, J., Willems, H., Torfs, R., et al. (2011). Impact of time–activity patterns on personal exposure to black carbon. *Atmospheric Environment, 45*(21), 3594–3602.

Dons, E., Panis, L. I., Van Poppel, M., Theunis, J., & Wets, G. (2012). Personal exposure to black carbon in transport micro-environments. *Atmospheric Environment, 55,* 392–398.

English, L. L., Munsmuir, D., Kumbakumba, E., Ansermino, J. M., Larson, C. P., Lester, R., et al. (2016). The Paediatric risk assessment (PARA) mobile app to reduce post discharge child mortality: design, usability, and feasibility for health care workers in Uganda. *JMIR mHealth and uHealth, 4*(1).

Fang, T. B., & Lu, Y. (2012). Personal real-time air pollution exposure assessment methods promoted by information technological advances. *Annals of GIS, 18*(4), 279–288.

Funk, S., Gilad, E., Watkins, C., & Jansen, V. A. (2009). The spread of awareness and its impact on epidemic outbreaks. *Proceedings of the National Academy of Sciences, 106*(16), 6872–6877.

Gerharz, L. E., Krüger, A., & Klemm, O. (2009). Applying indoor and outdoor modeling techniques to estimate individual exposure to PM2. 5 from personal GPS profiles and diaries: a pilot study. *Science of the Total Environment, 407*(18), 5184–5193.

Hess, G. R., Randolph, S. E., Arneberg, P., Chemini, C., Furlanello, C., Harwood, J., et al (2001). Spatial aspects of disease dynamics. In P. J. Hudson, A. P. Rizzoli, B. T. Grenfell, J. A. P. Heesterbeek, A. P. Dobson (Eds.), *The Ecology of Wildlife Diseases* (pp. 102–118). Oxford: Oxford University Press.

Kiss, I. Z. (2013). Incorporating human behaviour in epidemic dynamics: a modeling perspective. In *Modeling the Interplay between Human Behaviour and the Spread of Infectious Diseases*, (pp. 125–137). New York: Springer.

Kwan, M. P. (2009). From place-based to people-based exposure measures. *Social Science and Medicine, 69*(9), 1311–1313.

Lessler, J., Reich, N. G., & Cummings, D. A. (2009). Outbreak of 2009 pandemic influenza A (H1N1) at a New York City school. *New England Journal of Medicine, 361*(27), 2628–2636.

Leyk, S., Binder, C. R., & Nuckols, J. R. (2009). Spatial modeling of personalized exposure dynamics: the case of pesticide use in small-scale agricultural production landscapes of the developing world. *International Journal of Health Geographics, 8*(1), 1–17.

Liu, H. Y., Skjetne, E., & Kobernus, M. (2013). Mobile phone tracking: in support of modelling traffic-related air pollution contribution to individual exposure and its implications for public health impact assessment. *Environmental Health, 12*(1), 93.

Mao, L. (2014). Modeling triple-diffusions of infectious diseases, information, and preventive behaviors through a metropolitan social network—an agent-based simulation. *Applied Geography, 50,* 31–39.

Martin, C. K., Gilmore, A., Apolzan, J. W., Myers, C. A., Thomas, D. M., & Redman, L. M. (2016). Smartloss: a personalized mobile health intervention for weight management and health promotion. *JMIR mHealth and uHealth, 4*(1).

Meloni, S., Perra, N., Arenas, A., Gomez, S., Moreno, Y., & Vespignani, A. (2011). Modeling human mobility responses to the large-scale spreading of infectious diseases. *Scientific Reports, 1,* 62. https://doi.org/10.1038/srep00062.

Perchoux, C., Chaix, B., Cummins, S., & Kestens, Y. (2013). Conceptualization and measurement of environmental exposure in epidemiology: accounting for activity space related to daily mobility. *Health and Place, 21,* 86–93.

Perra, N., & Vespignani, A. (2013). Modeling contact and mobility based social response to the spread of infectious disease. In *Modeling the Interplay between Human Behaviour and the Spread of Infectious Diseases*, (pp. 103–123). New York: Springer.

Rothman, K. J., Greenland, S., & Lash, T. L. (2008). *Modern Epidemiology*. Lippincott Willians Wilkins.

Samanlioglu, F., Bilge, A. H., & Ergonul, O. (2012). A susceptible-exposed-infected-removed (SEIR) model for the 2009–2010 A/H1N1 epidemic in Istanbul. arXiv preprint arXiv: 1205–2497.

Sattenspiel, L. (2009). *The Geographic spread of infectious diseases*. Princeton University Press.

Setton, E., Marshall, J. D., Brauer, M., Lundquist, K. R., Hystad, P., Keller, P., et al. (2011). The impact of daily mobility on exposure to traffic-related air pollution and health effect estimates. *Journal of Exposure Science and Environmental Epidemiology, 21*(1), 42–48.

Stanton, M., Molineux, A., Mackenzie, C., & Kelly-Hope, L. (2016). Mobile technology for empowering health workers in underserved communities: new approaches to facilitate the elimination of neglected tropical diseases. *JMIR Public Health and Surveillance, 2*(1).

Steinle, S., Reis, S., & Sabel, C. E. (2011). Assessment of personal exposure to air pollutants in Scotland—an integrated approach using personal monitoring data. In *19th International Congress on Modelling and Simulation*, Perth, Ausrtalia, 12–16 December.

Steinle, S., Reis, S., Sabel, C. E., Semple, S., Twigg, M. M., Braban, C. F., Leeson, S. R., Heal, M. R., Harrison, D., Lin, C., Wu, H. (2015). Personal exposure monitoring of PM 2.5 in indoor and outdoor microenvironments. *Science of the Total Environment, 508,* 383–394.

Stoddard, S. T., Morrison, A. C., Vazquez-Prokopec, G. M., Paz Soldan, V., Kochel, T. J., Kitron, U., Elder, J. P., & Scott, T. W. (2009). The role of human movement in the transmission of vector-borne pathogens. *PLOS Neglected Tropical Diseases, 3*(7). https://doi.org/10.1371/journal.pntd.0000481.

Tsai, Y. S., Huang, C. Y., Wen, T. H., Sun, C. T., & Yen, M. Y. (2010). Integrating epidemic dynamics with daily commuting networks: buildings a multilayer framework to assess influenza A(H1N1) intervention policies. *Simulation: Transactions of the Society for Modeling and Simulation International, 87*(5), 385–405.

Wang, S. W., Tang, X., Fan, Z. H., Wu, X., Lioy, P. J., & Georgopoulos, P. G. (2009). Modeling of personal exposures to ambient air toxics in Camden, New Jersey: an evaluation study. *Journal of the Air Waste Management Association, 59*(6), 733–746.

Wang, B., Cao, L., Suzuki, H., & Aihara, K. (2012). Safety-information-driven human mobility patterns with metapopulation epidemic dynamics. *Scientific Reports, 2*(887). https://doi.org/10.1038/srep00887.

Weis, B. K., Balshaw, D., Barr, J. R., Brown, D., Ellisman, M., Lioy, P., et al. (2005). Personalized exposure assessment: promising approaches for human environmental health research. *Environmental Health Perspectives, 113*(7), 840–848.

Wu, Q., Fu, X., Small, M., & Xu, X. J. (2012). The impact of awareness on epidemic spreading in networks. *Chaos: An Interdisciplinary Journal of Nonlinear Science, 22*(1). https://doi.org/10.1063/1.3673573.

Zhao, S., Wu, J., & Ben-Arieh, D. (2014). Modeling infection spread and behavioral change using spatial games. *Health System, 4*(1), 41–53.

Author Biographies

Tzai-Hung Wen (Ph.D., National Taiwan University) Professor in the Department of Geography at National Taiwan University. His research interests include using multidimensional spatial-temporal data from mobile geo-sensors, social media and open data platforms for exploring geospatial process in human environment. Prof. Wen's studies focus on developing geo-computational methods for quantifying the role of human mobility in infectious disease diffusion.

Ching-Shun Hsu (M.Sc., National Taiwan University) Doctoral student in the Department of Geography at National Taiwan University. His research interests include spatial-temporal analysis and infectious disease modeling. Mr. Hsu's studies focus on integrating routine disease surveillance with patient flow data for developing epidemic early-warning framework.

Chih-Hong Sun (Ph.D., University of Georgia) Professor in the Department of Geography at National Taiwan University. He is also a former President of the Asia GIS Association. His research interests include geographic information science (GIScience), ontology-based spatial decision support system (SDSS) and smart city. Prof. Sun's studies focus on employing the Open Semantic Framework (OSF) to integrate various kind of structured, semi-structured, and unstructured data for developing Smart Earth Collaboration Platform (SECP).

Joe-Air Jiang (Ph.D., National Taiwan University) Distinguished Professor in the Department of Bio-Industrial Mechatronics Engineering at National Taiwan University and the Chairman of the Taiwan Institute of Biological Mechatronics (TIBM). His research interests include wireless sensor network (WSN), Internet of Things (IoT), automatic control and electronics. Prof. Jiang's studies focus on designing software and hardware of WSN, which integrates distributed sensing capability, real-time data analysis, and remote surveillance with ubiquitous Internet connection for ecological applications.

Jehn-Yih Juang (Ph.D., Duke University) Associate Professor in the Department of Geography at National Taiwan University. His research interests include surface-atmosphere interactions, environmental measurement and modeling, especially air pollution statistical and numerical modeling. Dr. Juang' studies focus on these related environmental topics in Taiwan and the vicinity in East Asia.

Chapter 8
An Exploratory Analysis of the Effects of Spatial and Temporal Scale and Transportation Mode on Anonymity in Human Mobility Trajectories

Jennifer A. Miller and Brendan Hoover

8.1 Background

Advancements in location-acquisition technologies such as global positioning systems (GPS), radio frequency identification (RFID), cellular phone networks, and WiFi hotspots have resulted in significant increases in the availability of highly accurate data on moving objects, with unprecedented high spatial and temporal resolution. These location data are often studied as 'trajectories', comprised of a series of time-stamped sequential locations. Moving objects of interest have extended beyond the traditional scale associated with people, animals, and vehicles to include weather events such as hurricanes (Dodge et al. 2012) and eye tracking, where gaze trajectories are compared with computer mouse movement to study human-computer interaction (Demšar et al. 2015).

As a result of the increasingly wide range of types of moving objects studied, several different interdisciplinary communities are now focusing on issues associated with collecting, managing, visualizing, and analyzing spatio-temporal data associated with moving objects. Originating from the relatively long history of animal tracking and telemetry studies, 'movement ecology' has become a rapidly growing subfield in ecology focused on understanding the "causes, mechanisms, and spatiotemporal patterns of (organismal) movement and their role in various ecological and evolutionary processes" (Nathan et al. 2008: 19,052). Within geographic information science (GIScience), 'computational movement analysis' has recently emerged as a subfield that focuses on the development and application of computational techniques for collecting, managing, and analyzing movement data in order to better understand the processes that are associated with them (Gudmundssen et al. 2012). 'Trajectory data mining' harnesses new computing

J. A. Miller (✉) · B. Hoover
Department of Geography and the Environment, The University of Texas
at Austin, Austin, TX, USA
e-mail: jennifer.miller@austin.utexas.edu

© Springer International Publishing AG, part of Springer Nature 2018   149
S.-L. Shaw and D. Sui (eds.), *Human Dynamics Research in Smart
and Connected Communities*, Human Dynamics in Smart Cities,
https://doi.org/10.1007/978-3-319-73247-3_8

technologies to discover knowledge from trajectory data (Zheng 2015) with applications mainly in location-based social networks, transportation systems, and urban computing.

Depending upon the technology used to collect the data, the location information can be represented by precise latitude and longitude coordinates (e.g., GPS data from a smartphone or other device) or the catchment area of a single cellular tower (e.g. call detail records (CDR) from cellular phones). These relatively low cost location data have been used to explore human mobility patterns related to, for example, urban planning (Steenbruggen et al. 2015), transportation infrastructure (Wu et al. 2013), disaster planning/evacuation strategies (Ghurye et al. 2016), potential disease spread (Oliver et al. 2015), and many other applications (see review by Becker et al. 2013). Deville et al. (2014) introduced a framework for using mobile phone data to calculate temporally explicit population data at the spatial resolution of cellular tower service areas in order to supplement census data and better understand human dynamics. Mobile phone data have also been used to represent spatiotemporal human mobility dynamics in the context of the spread of diseases such as malaria in Kenya (Wesolowski et al. 2012) and cholera in Senegal (Finger et al. 2016). The spatial resolution of CDR data varies as a function of population density. In their seminal study in an unnamed Western European country, de Montjoye et al. (2013) reported CDR catchment areas ranging from 0.15 km^2 in cities to 15 km^2 in rural areas.

Higher precision GPS data, typically collected with user- or vehicle-carried smartphones, have been used to study route choice behavior (Huang and Levinson 2015), the effects of built environment on physical activity and health (Collins et al. 2012; Carlson et al. 2015), to determine the risk of cycling injury (Strauss et al. 2015), to detect travel model (Xiao et al. 2015), and to provide social itinerary recommendations (Yoon et al. 2011). Sila-Nowicka et al. (2016) explored contextual information linked to GPS locations in order to identify "third places" beyond home and work.

8.1.1 Location Privacy

While studies using mobile phone and GPS data have made important contributions to a better understanding of human mobility and spatiotemporal dynamics in general, there are significant issues associated with the distribution or availability of these data. As often happens with technological advancements, the collection of these data has preceded extensive study on how and what they can (or should) be used for, as well as the privacy implications associated with them.

In particular, there are important privacy issues associated with location or mobility data that can be traced to a single or very few individuals that are often overlooked. Wernke et al. (2012) classify potential location privacy attacks based on the information possessed by the adversary: single or multiple locations, contextual information, and historical information. Ma et al. (2013) provide defensive approaches for scenarios in which an adversary is passive and is given individual mobility data as well as scenarios in which an adversary is active and physically collects contextual and/or ancillary data along with the mobility data. In addition to what might be considered superficially harmless privacy violations, an adversary could use location data for malicious purposes such as stalking or opportunistic criminal activities. Locations or movement patterns could also be used to make inferences that have potentially negative implications, for example, repeated visits to a medical clinic may be a concern for a prospective employer.

Measures that have been implemented to preserve privacy have been shown to be superficial or ineffective. Location data are often released after they have been 'anonymized'—which means that the trajectory has been stripped of any identifying information such as name, address, and phone number. However, as Golle and Partridge (2009) note, anonymity is "a useful but imperfect tool for preserving location privacy" (p. 390). Personal points of interest (home, work) can still be identified by mining trajectory data for movement patterns, and these points of interest are often associated with unique individuals.

Using an extensive dataset of home and work locations in the U.S., Golle and Partridge (2009) showed that at the spatial scale of a census block, the pair of home/work locations is unique for a majority of the working population. At the scale of census tracts, the pair of locations was uniquely identifying for only 5% of the working population but at the much coarser county scale, the 44% of workers who live and work in different counties are considerably more vulnerable to de-anonymization. Zang and Bolot (2011) used anonymized CDR from 25 million individuals across the U.S. to determine the "top N" locations at which calls were recorded for each of three months. They found that when $N = 2$ (typically corresponding to work and home locations), up to 35% of the individuals could be uniquely identified. When $N = 3$ (they suggested the 3rd location typically represented a school or shopping related location), 50% could be uniquely identified.

It should be emphasized here that the concept of unicity or the ability to uniquely identify a movement trajectory based on a small subset of locations of which it is comprised is not equivalent to de-anonymization, but it is a requisite first step. Quantifying the uniqueness of locations through which an individual moves is necessary to better understand privacy implications associated with increasingly available human mobility datasets.

In one of the first studies to address quantification of unicity of individual trajectories, de Montjoye et al. (2013) used fifteen months of anonymized mobile phone data (CDR) for 1.5 million individuals in a western European country and found that four randomly selected spatiotemporal points were sufficient to uniquely identify 95% of the individuals. Perhaps more troubling, they found that over 50% of individuals were uniquely identifiable from just two randomly selected locations (typically also corresponding to home and work). Song et al. (2014) found similar results with a dataset of one week of mobility data for 1.14 million people (total 56 million records): 60% of the trajectories were unique using just two random points.

Due to data availability, most of the previous work on measuring 'unicity' or the uniqueness of movement traces or trajectories has been with much coarser scaled cell phone data. However, even relatively coarse spatial resolution location data such as that associated with call detail records (CDR), where 'location' is an area defined by its proximity to a specific cell phone tower, can be used to uniquely identify an individual. Locations of cell phone towers or antennae are based on population density and the area associated with each one varies considerably. In their study in a small (unnamed) European country, de Montjoye et al. (2013) found that the reception or catchment area for an antenna ranged from 0.15 km^2 in urban areas to 15 km^2 in rural areas.

It is important to note that uniqueness does not equate to re-identifiability and the objectives of these studies were to examine how unique individual trajectories were, not to actually de-anonymize them or re-attach an individual's information to a unique trajectory. However, the ability to measure uniqueness of locations on a trajectory is an important prerequisite for re-identification (which would involve correlation with an ancillary dataset) and therefore, represents a potential threat to individual privacy.

8.1.2 Measuring Unicity

While a single widely used or standard measure of unicity has not yet emerged, unicity has been measured in different ways, depending on characteristics of the mobility dataset and research objectives. In studies similar to the research presented here, unicity has typically been quantified by comparing a randomly selected subset of points (either location or location + time) to points in the mobility trajectories; unicity would be high if the subset of points matches very few other trajectories. Unicity would be low (and location privacy less problematic) if the subset of points matched many other trajectories. de Montjoye et al. (2013) measured 'unicity' as the percentage of 2500 random traces that were unique given p random points (p ranged from 2 to 5). Song et al. (2014) defined uniqueness of trajectories as the percentage of all available trajectories that were uniquely associated with p random points, which they varied from 2 to 6. While anonymity (or lack thereof) has been studied with CDR data, as the previous examples show, it has not yet been more extensively addressed with finer spatiotemporal resolution available as GPS locations from, e.g., smartphones (but see Rossi et al. 2015). These datasets could potentially be far more unique and therefore more difficult to anonymize.

In addition to location and location + time, Rossi et al. (2015) also tested how additional movement information derived from three different published mobility datasets could be uniquely associated with individual trajectories. They calculated distance traveled, average speed, and average angle of travel for a specific time window and measured unicity as the average uniqueness over the whole dataset using 1000 subsets of points (number of points ranged from 1 to 5) per individual. They found that direction was the most unique movement parameter for all three datasets, with five points able to uniquely identify 95% of the users.

The degree of uniqueness of trajectories can vary as a function of factors such as typical commuting patterns, transportation modes, and geographical region (which also affects commuting patterns and transportation modes). While measurement precision and geographic scale have been varied in order to assess their effects on unicity of a mobility dataset, the influence of these other factors has rarely been examined. Information on transportation modes associated with trajectory segments has only recently been provided as a component of some mobility datasets, although previous research has focused on inferring likely transportation modes based on movement characteristics analyzed from the locations (Lin and Hsu 2014; Zheng et al. 2010). Sila-Nowicka and Thakuriah (2016) used travel diaries and mobility data for 358 users in Glasgow, Scotland to compare spatial patterns between the original mobility data and generalized data that resulted from kernel density estimation for four different travel modes (driving, walking, train, bus). They calculated Pearson's correlation coefficient for the original and generalized data and the values ranged from 1.0 for train to 0.852 for walking. No previous studies explicitly examined how different transportation modes affected unicity.

The importance of quantifying the 'anonymity' of a database has been a research focus in information sciences far longer than the relatively new issue of unicity and there are several widely accepted methods. The most commonly used method of k-anonymity was introduced by Sweeney (2002) as a measure to increase anonymity for non-spatial databases. When applied to spatial databases, it ensures that any set of records (locations) for an individual is at least the same as $k - 1$ individuals. Generally, $k = 2$, ensuring that at least two trajectories are equivalent, but as k increases, so too does the anonymity. Extensions of k-anonymity include l-diversity and t-closeness (Li et al. 2007).

These measures are generally used to manage trajectory datasets (e.g., data would be manipulated so that the level of anonymity reached the reported k level), but in order to quantify the actual level of anonymity of trajectory datasets, a rigorous analysis comparing random points from each trajectory to all other trajectories still needs to be conducted. With trajectory datasets now available at one second temporal resolution, the volume of these data can result in computationally intensive analysis.

In addition to removing any identifying information, 'cloaking' or other obfuscation techniques have been used to add noise to or reduce the precision of location data associated with mobility traces (Gambs et al. 2010; Ma et al. 2013). While there has been extensive research on approaches to preserve location privacy (see Kar et al. 2013; Seidl et al. 2016) most of the methods result in significant loss of information and none has been considered to be broadly successful (Narayanan and Felten 2014).

The research presented here explores issues related to privacy and identity associated with more recently available high resolution GPS location data. We quantified the unicity of GPS movement trajectories testing the effect of spatial resolution and temporal resolution. In addition to location, we explored how effective derived movement parameters such as direction could be for uniquely identifying a trajectory. We also calculated unicity for different user-labelled transportation modes and explored how it is affected by spatial resolution.

8.2 Data

We explored these issues using GeoLife Trajectories (Zheng et al. 2008a, b, 2010), a well-known mobility data of individuals in Beijing, China collected by Microsoft Asia. This is an extremely dense dataset, with temporal resolution of ~ 1–5 s and spatial resolution of ~ 5–10 m. For the first part of our analysis, we used only one year of data (January 2009–December 2009) and the spatial extent of Beijing (39.6°–40.2°N latitude and 116°–116.8°E longitudes) to remove users who traveled outside of the city during this time period. This resulted in 71 users who had a total of 7243 daily trajectories (number of locations visited within trajectories varied but the mean was 1600). Here we use an individual's daily trajectory (where a day is considered to begin and end at midnight) as the basic trajectory unit; each begins and ends at a time that is dependent on the individual's daily activity.

The second part of our analysis focused on how unicity varied as a function of transportation mode. Individual-labelled transportation modes (bus, subway, train, taxi, car, walk) were only available for 69 individuals in the GeoLife dataset, so we did not confine these trajectories to a certain year and therefore did not include temporal information in the unicity analysis. Transportation mode was attached to the appropriate section of an individual's trajectory and we considered trajectories to be separate if they were at least fifteen minutes apart even if they involved the same transportation mode.

8.3 Methods

The basis of our unicity test involved extracting 500 sets of points of size n ($n = 2, 3, 4,$ and 5 points) from each user and counting how many other trajectories contain those points. The percentage of 500 sets of points that matched only one trajectory was calculated and this was done for each of the 71 users for the four different point sizes (n = 2, 3, 4, and 5). Our measure of unicity, u, was the percentage of 500 random points of size n that are contained in only one trajectory averaged across all 71 users. A unicity value close to 100 indicates a highly unique trajectory that could theoretically be de-anonymized, or re-connected with identifying user information more easily; a low unicity value suggests that the random set of points are contained in several different trajectories and therefore would make de-anonymizing trajectories far more challenging. The amount of information from each point was varied— we used just location (x and y), location + time (x, y, and t), and the direction (the absolute angle for point i is measured between the x direction and the step built by relocations i and i + 1).

The original latitude and longitude coordinates for these locations had a spatial precision of six decimal places (~ 0.1 m). In order to test how spatial and temporal resolution affected measurement of unicity, the geographic coordinates were coarsened first to four decimal places (~ 10 m) and the temporal resolution was coarsened to 30 s, then further coarsened to three decimal places (100 m) and 60 s. Additionally, the precision of the absolute angle measure was decreased from the original (five decimal places) to three decimal places.

For the transportation mode analysis, unicity was calculated the same way, except the n points were compared only to points labelled with the same transportation mode. The same two levels of coarsening were applied to the geographic coordinates, but temporal information was not used here in order to provide a more general location comparison. The use of different transportation modes varied widely. The 69 individuals collectively used a total of 12,291 transportation-labelled trajectory segments, but the number of labelled trajectories per individual ranged from 1 to ~ 2800 (i.e., some individuals had only one transportation-labelled daily trajectory, while others had several thousand). The mean use of each transportation mode per individual, along with minimum and maximum, is given in Fig. 8.1 (bars) along with the mean percentage of total trajectory sections per transportation mode (line). Walking was the most frequently reported transportation mode (52% of all trajectory segments) while train was the least frequently used (2%). Cars represented only about 8% of all trajectory segments, but the mean percentage of use across each individual's trajectory segments was 15%, with a minimum of 0% and maximum of 100% (i.e., at the extremes, some individuals never used cars for transportation, while cars were the exclusive transportation model for other individuals).

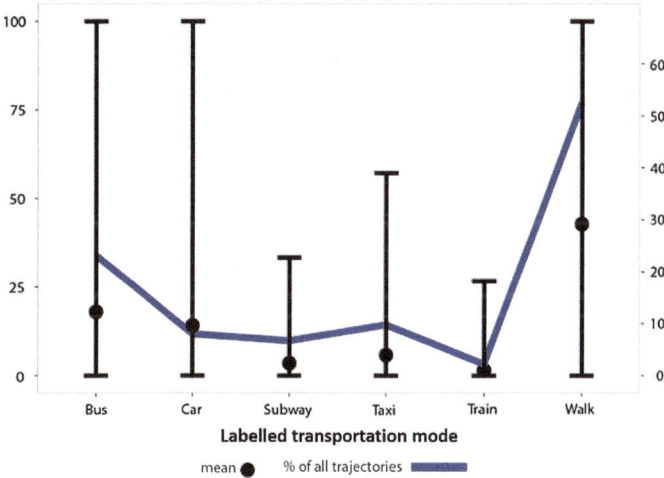

Fig. 8.1 The mean, minimum, and maximum of the percentage of each individual's labelled trajectory segments that was associated with each transportation mode (bars, left axis); and the percentage of total labelled trajectory segments associated with each transportation mode (line, right axis)

8.4   Results

The mean unicity values associated with the size of each random point set and level of coarsening for location, location + time, and direction (absolute angle) are plotted in Fig. 8.2. In general, the locations on a trajectory were highly unique. 90% of the random sets of just two points composed of only location (no timestamp) were associated with only one trajectory. Adding the timestamp increased the unicity of two points to 97%. When five points with location and timestamp were used, the unicity increased to almost 99%. The implications for location privacy are alarming, as these were randomly selected locations and not the 'most visited' that might be associated with potentially more unique work-home pairs of locations. Somewhat surprisingly, the angle of movement alone also had high unicity—when the angles of four points were tested, the unicity ($u = 66\%$) was similar or greater than the unicity of location for CDR using two points as found in de Montjoye et al. (2013) and Song et al. (2014). Five angle values could uniquely identify a trajectory 73% of the time, although coarsening the precision of the angle measurements had a much more negative effect on unicity. More research is needed to address the unicity of derived movement parameters separated from actual locations as a potential privacy issue.

When just two points (no timestamp) were used at the first level of coarsening (spatial precision reduced tenfold to ~ 10 m), unicity was still almost 69%; when the coarsened time (~ 30 s) information was added to location, the unicity was similar to the original resolution (88%).

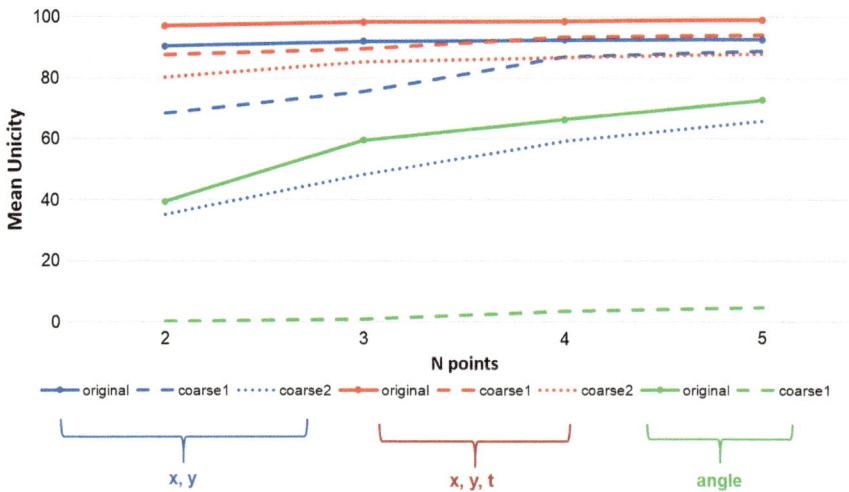

Fig. 8.2 Mean unicity for different types of information (location in blue, location + time in red, and absolute angle (direction) in green) and different measurement precision (solid line is original precision, dashed line is first level of coarsening, and dotted line is second level of coarsening). The number of sample points compared is on the x-axis

The final level of coarsening decreased the spatial resolution of an x/y pair to ∼100 m and the temporal resolution was coarsened to one minute. The spatial resolution here was closer to the resolution of the antenna reception areas used in the de Montjoye et al. (2013) paper (where spatial resolution ranged from 115 m to 15 km), but the coarsened temporal resolution was still much more precise than the one used in the CDR studies. As a result, using location + time for just two points resulted in a high unicity (mean 80.3%), while five points increased the mean unicity to almost 88%. Using just location (no timestamp), the unicity degraded to 32% for two points and 66% for five points.

Using four or five points, the first level of coarsened time + location had similar unicity to the original precision location coordinates (93–94%), while coarsest time + location was similar to the first level of coarsened location coordinates (86–87%). Only one level of coarsening was used for direction, as unicity of the absolute angle degraded substantially—unicity was only 5% when a set of five points was used.

Unicity of trajectories (location only) associated with different transportation modes is plotted in Fig. 8.3. We don't include the results from trajectories labelled with train, as it was only used by 20 individuals and represented only 2% of all labelled trajectory segments. Walking had the highest unicity for all four point sample sizes (u ranged from 86% for two points to 93% for five points). Subway

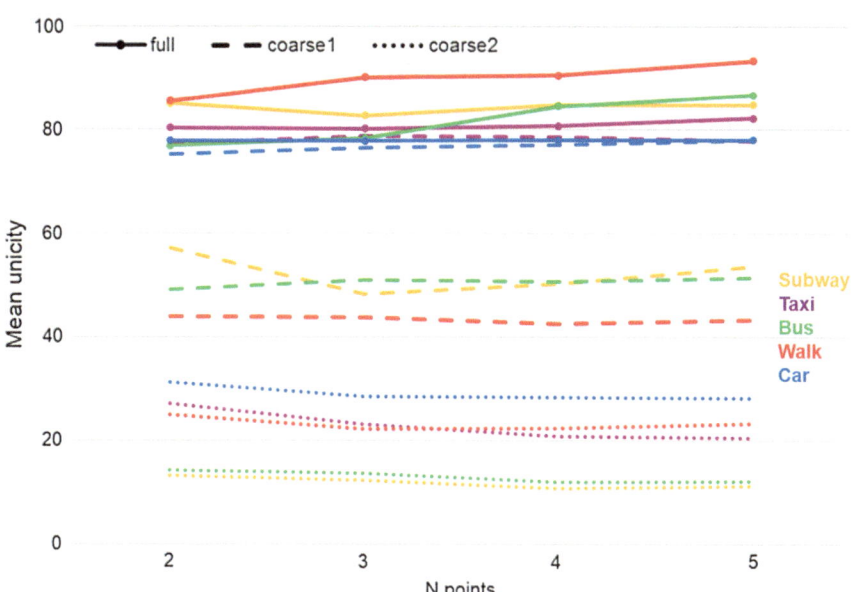

Fig. 8.3 Mean unicity for five transportation modes (subway is yellow, taxi is purple, bus is green, walk is red, and car is blue) and three measurement levels (solid line is original precision, dashed line is first level of coarsening, and dotted line is second level of coarsening). The number of sample points compared is on the x-axis

mode was the next most unique transportation mode, but had slightly higher unicity with two sample points (85%) than for three points (83%) and five points only increased unicity to 85%. Trajectory sections that involved taxis and cars had the most constant unicity across all four sample sizes, ranging from $u = 80\%$ (taxis) and 77% (cars) for two points to $u = 82\%$ (taxis) and 78% (cars) for five points. These transportation modes were also relatively less frequently used (10% of all trajectory segments involved taxis while 8% involved cars). Unicity associated with buses increased from 77% with two points to 87% with five points.

The spatial resolution of the location coordinates was coarsened to the same two levels as described above (10 and 100 m) in order to assess how unicity associated with transportation modes was affected. Unicity for taxis and cars was only slightly decreased at the first level of coarsening and unicity for both modes was fairly constant across all sample sizes. Unicity did decrease substantially for taxis and cars at the second level of coarsening, although unicity for cars remained higher than for taxis (28–30% and 27–20%, for cars and taxis respectively). Unicity also decreased marginally as the number of sample points increased for taxis and cars, although this could result from their relative scarcity in the dataset.

Unicity for subways, buses, and walking decreased markedly (u was less than 60%) with the first level of coarsening. While buses and walking had similar unicity across all sample sizes, subways had the highest unicity with just two sample points (57%). At the coarsest level, unicity dropped considerably for all five transportation modes, ranging from 31% for cars with two sample points to 12% with five points for buses and subways.

8.5   Conclusion

With the much higher precision and spatial resolution of GPS data currently available, two x/y locations are sufficient to be uniquely associated with a single trajectory 90% of the time, adding the timestamp matches a single trajectory 97% of the time. The three pieces of information—location + time—are so specific that increasing the number of points to match to five increases the unicity very little because it is already so high using just two points. The first level of coarsening for location + time (~ 10 m spatial, 30 s temporal) has similar unicity to the original resolution for just location coordinates, and when four or five points are used, the coarsened location + time has slightly higher mean unicity. The location coordinates (no timestamp) show the greatest increase in unicity when additional points are used for matching. This suggests that there is a trade-off between location resolution and amount of information (location points) available.

The relatively unique signature of derived movement information alone highlights potential location privacy issues even when location information (x/y) have been removed from the dataset. Movement parameters such as speed, angle, and step length have rarely been tested as potential identifiers of trajectories, but the case study here focusing on absolute angle highlights their potential importance.

Five absolute angle data points were uniquely associated with a single trajectory 72% of the time. This suggests that individual movement, irrespective of absolute geographic location, can be identifiable with a sufficient level of precision of angle measurements and data points. Future work should focus specifically on how movement parameters could be used singly or together to identify a trajectory.

It was not surprising that unicity varies with transportation mode. Among the five transportation modes used here, walking is the least constricted while subway is likely the most constricted. Consequently, walking was the most unique travel mode at the original precision, although unicity was greatly decreased as precision decreased. This is particularly disconcerting given that walking individuals are perhaps the most vulnerable in the broadest sense of potential privacy attacks. All five transportation modes had unicity values of 78% or higher at original precision for as few as two points. The effect of transportation mode and unicity would also likely vary based on spatial characteristics of a city (ex., extent of sprawl), population characteristics of a city, transportation infrastructure, climate and other factors. Unicity would also be expected to be dependent on the amount of time an individual spends on a public transportation mode—shorter routes would presumably have lower unicity while longer routes might have higher unicity (with exceptions related to well-traveled but more distant destinations such as an airport). The unicity associated with taxi, car, bus, and subway transportation modes was similar at the original resolution, although there were far more occurrences of bus travel in the data. However, at the first level of coarsening, taxi and car mode continued to have high unicity, even higher than walking. This suggests that the spatial scale at which roads are resolved contributes to higher unicity. More research is needed to better understand how unicity varies with transportation mode based on user-provided travel information.

This unicity study has particularly important implications for privacy and the increasing availability of 'anonymized' trajectory datasets. This is one of the few studies to explore unicity with higher resolution GPS data and it should be troubling how unique a set of two location points can be. Coarsening the spatial and temporal resolution reduces the unicity, but five points with x, y coordinates at the coarsest resolution tested here were still uniquely associated with a single trajectory more than 60% of the time. Our results also show an increase in unicity when more than two points are used, presumably the "third place" (see Sila-Nowicka et al. 2016) after points representing e.g., home and work can be an important determinant of unicity. This effect persists with scale coarsening and when just angle of movement is used. We detected a few instances where trajectories were duplicated in this dataset, which could result in conservative estimates of unicity. Conversely, the calculation of unicity will be dependent upon how many individual trajectories are used in the comparison—as we use trajectories for ~70 individuals, unicity could potentially be overstated here. More research is needed on how calculation of unicity is affected by dataset characteristics like number of individuals and trajectories, geographic region, infrastructure, and social factors that affect movement patterns.

It is important to note here that the focus of this study was not to re-attach user information to trajectories, it was just to examine how unique trajectories were based on different types and levels of information stored with or derived from them. Problems with anonymizing published mobility datasets have already been highlighted using the relatively coarse spatial resolution of call detail records. These privacy issues will only be exacerbated as higher quality GPS location datasets become increasingly available. The dataset used here represents trajectories for 71 and 69 individuals (transportation mode data), which may be too small to make inferences about unicity. GeoLife was not originally collected for studies focused on location privacy, although it has been used in similar studies (Rossi et al. 2015).

In addition to measuring unicity of location and location + time, further study is needed on how different factors such as transportation mode and movement parameters affect unicity as well as potential implications involving linked datasets from social surveys. While each of the issues addressed here focuses on a single dataset for the case study, we would expect the results to be broadly applicable to other similar mobility datasets.

Acknowledgments This research was funded in part by the Center for Identity at The University of Texas.

References

Becker, R., Cáceres, R., Hanson, K., Isaacman, S., Loh, J. M., Martonosi, M., et al. (2013). Human mobility characterization from cellular network data. *Communications of the ACM, 56,* 74–82. https://doi.org/10.1145/2398356.2398375.

Carlson, J. A., Saelens, B. E., Kerr, J., Schipperijn, J., Conway, T. L., Frank, L. D., et al. (2015). Association between neighborhood walkability and GPS-measured walking, bicycling and vehicle time in adolescents. *Health & Place, 32,* 1–7. https://doi.org/10.1016/j.healthplace. 2014.12.008.

Collins, P., Al-Nakeeb, Y., Nevill, A., & Lyons, M. (2012). The impact of the built environment on young people's physical activity patterns: A suburban-rural comparison using GPS. *International Journal of Environmental Research and Public Health, 9,* 3030–3050. https:// doi.org/10.3390/ijerph9093030.

Demšar, U., Buchin, K., Cagnacci, F., Safi, K., Speckmann, B., Van de Weghe, N., et al. (2015). Analysis and visualisation of movement: An interdisciplinary review. *Movement Ecology, 3,* 5. https://doi.org/10.1186/s40462-015-0032-y.

de Montjoye, Y.-A., Hidalgo, C. A., Verleysen, M., & Blondel, V. D. (2013). Unique in the crowd: The privacy bounds of human mobility. *Scientific Reports, 3,* 1376. https://doi.org/10. 1038/srep01376.

Deville, P., Linard, C., Martin, S., Gilbert, M., Stevens, F. R., Gaughan, A. E., et al. (2014). Dynamic population mapping using mobile phone data. *PNAS, 111,* 15888–15893. https://doi. org/10.1073/pnas.1408439111.

Dodge, S., Laube, P., & Weibel, R. (2012). Movement similarity assessment using symbolic representation of trajectories. *International Journal of Geographical Information Science, 26,* 1563–1588. https://doi.org/10.1080/13658816.2011.630003.

Finger, F., Genolet, T., Mari, L., de Magny, G. C., Manga, N. M., Rinaldo, A., et al. (2016). Mobile phone data highlights the role of mass gatherings in the spreading of cholera outbreaks. *Proc Natl Acad Sci U S A, 113*, 6421–6426. https://doi.org/10.1073/pnas.1522305113.

Gambs, S., Killijian, M.-O., del Prado Cortez, M. (2014). De-anonymization attack on geolocated data. *Journal of Computer and System Sciences, Special Issue on Theory and Applications in Parallel and Distributed Computing Systems, 80*, 1597–1614. https://doi.org/10.1016/j.jcss. 2014.04.024.

Gambs, S., Killijian, M.O., & del Prado Cortez, M.N. (2010). November. Show me how you move and I will tell you who you are. In *Proceedings of the 3rd ACM SIGSPATIAL International Workshop on Security and Privacy in GIS and LBS* (pp. 34–41). ACM.

Ghurye, J., Krings, G., & Frias-Martinez, V. (2016, June). A framework to model human behavior at large scale during natural disasters. In *2016 17th IEEE International Conference on Mobile Data Management (MDM)*, (Vol. 1, pp. 18–27). USA: IEEE.

Golle, P., & Partridge, K. (2009). On the Anonymity of Home/Work Location Pairs. In *Presented at the Proceedings of the 7th International Conference on Pervasive Computing* (pp. 390–397), Berlin: Springer. https://doi.org/10.1007/978-3-642-01516-8_26.

Gudmundsson, J., Laube, P., & Wolle, T. (2012). Computational movement analysis. In W. Kresse & D. M. Danko (Eds.), *Springer handbook of geographic information* (pp. 423–438). Berlin, Heidelberg: Springer.

Huang, A., & Levinson, D. (2015). Axis of travel: Modeling non-work destination choice with GPS data. *Transportation Research Part C: Emerging Technologies, Big Data in Transportation and Traffic Engineering 58, Part B*, 208–223. https://doi.org/10.1016/j.trc. 2015.03.022.

Kar, B., Crowsey, R. C., & Zale, J. J. (2013). The myth of location privacy in the United States: Surveyed attitude versus current practices. *The Professional Geographer, 65*, 47–64. https://doi.org/10.1080/00330124.2012.658725.

Li, N., Li, T., & Venkatasubramanian, S. (2007). t-closeness: Privacy Beyond k-Anonymity and l-Diversity, In *2007 IEEE 23rd International Conference on Data Engineering. Presented at the 2007 IEEE 23rd International Conference on Data Engineering* (pp. 106–115). https://doi.org/10.1109/icde.2007.367856.

Lin, M., & Hsu, W.-J. (2014). Mining GPS data for mobility patterns: A survey. *Pervasive and Mobile Computing, 12*, 1–16. https://doi.org/10.1016/j.pmcj.2013.06.005.

Ma, C. Y. T., Yau, D. K. Y., Yip, N. K., & Rao, N. S. V. (2013). Privacy vulnerability of published anonymous mobility traces. *IEEE/ACM Transactions on Networking, 21*, 720–733. https://doi.org/10.1109/TNET.2012.2208983.

Narayanan, A., & Felten, E. W. (2014). No silver bullet: De-identification still doesn't work. *White Paper*, July.

Nathan, R., Getz, W. M., Revilla, E., Holyoak, M., Kadmon, R., Saltz, D., et al. (2008). A movement ecology paradigm for unifying organismal movement research. *PNAS, 105*, 19052–19059. https://doi.org/10.1073/pnas.0800375105.

Oliver, N., Matic, A., & Frias-Martinez, E. (2015). Mobile network data for public health: Opportunities and challenges. *Frontiers in Public Health, 3*.

Rossi, L., Walker, J., & Musolesi, M. (2015). Spatio-temporal techniques for user identification by means of GPS mobility data. *EPJ Data Science, 4*, 11. https://doi.org/10.1140/epjds/s13688-015-0049-x.

Seidl, D. E., Jankowski, P., & Tsou, M.-H. (2016). Privacy and spatial pattern preservation in masked GPS trajectory data. *International Journal of Geographical Information Science, 30*, 785–800. https://doi.org/10.1080/13658816.2015.1101767.

Sila-Nowicka, K., & Thakuriah, P. (2016). The trade-off between privacy and geographic data resolution. A case of GPS trajectories combined with the social survey results [WWW Document]. *ISPRS—International Archives of the Photogrammetry, Remote Sensing and Spatial Information Sciences*. URL http://www.isprs.org/publications/archives.aspx (accessed 10.16.16).

Siła-Nowicka, K., Vandrol, J., Oshan, T., Long, J. A., Demšar, U., & Fotheringham, A. S. (2016). Analysis of human mobility patterns from GPS trajectories and contextual information. *International Journal of Geographical Information Science, 30*(5), 881–906.

Song, Y., Dahlmeier, D., & Bressan, S. (2014). Not so unique in the crowd: A simple and effective algorithm for anonymizing location data. *ResearchGate, 1225,* 19–24.

Steenbruggen, J., Tranos, E., & Nijkamp, P. (2015). Data from mobile phone operators: A tool for smarter cities? *Telecommunications Policy, 39*(3), 335–346.

Strauss, J., Miranda-Moreno, L. F., & Morency, P. (2015). Mapping cyclist activity and injury risk in a network combining smartphone GPS data and bicycle counts. *Accident Analysis and Prevention, 83,* 132–142. https://doi.org/10.1016/j.aap.2015.07.014.

Sweeney, L. (2002). K-anonymity: A model for protecting privacy. *International Journal of Uncertainty, Fuzziness and Knowledge-Based Systems, 10,* 557–570. https://doi.org/10.1142/S0218488502001648.

Wernke, M., Skvortsov, P., Dürr, F., & Rothermel, K. (2012). A classification of location privacy attacks and approaches. *Personal and Ubiquitous Computing, 18,* 163–175. https://doi.org/10.1007/s00779-012-0633-z.

Wesolowski, A., Eagle, N., Tatem, A. J., Smith, D. L., Noor, A. M., Snow, R. W., et al. (2012). Quantifying the impact of human mobility on malaria. *Science, 338,* 267–270. https://doi.org/10.1126/science.1223467.

Wu, W., Cheu, E. Y., Feng, Y., Le, D. N., Yap, G. E., & Li, X. (2013). Studying intercity travels and traffic using cellular network data. Conference on *Mobile Phone Data for Development: Net Mob 2013*. Retrieved from http://perso.uclouvain.be/vincent.blondel/netmob/2013/D4D-book.pdf

Xiao, G., Juan, Z., & Zhang, C. (2015). Travel mode detection based on GPS track data and Bayesian networks. *Computers, Environment and Urban Systems, 54,* 14–22. https://doi.org/10.1016/j.compenvurbsys.2015.05.005.

Yoon, H., Zheng, Y., Xie, X., & Woo, W. (2011). Social itinerary recommendation from user-generated digital trails. *Personal and Ubiquitous Computing, 16,* 469–484. https://doi.org/10.1007/s00779-011-0419-8.

Zang, H., & Bolot, J. (2011). Anonymization of location data does not work: A large-scale measurement study. In *Presented at the Proceedings of the 17th annual international conference on Mobile computing and networking*, (pp. 145–156). USA: ACM. https://doi.org/10.1145/2030613.2030630.

Zheng, Y. (2015). Trajectory data mining: An overview. *ACM Transactions on Intelligent Systems and Technology, 6*(29):1–29:41. https://doi.org/10.1145/2743025.

Zheng, Y., Li, Q., Chen, Y., Xie, X., & Ma, W. -Y. (2008a). Understanding mobility based on GPS data. In *Presented at the Proceedings of the 10th international conference on Ubiquitous computing* (pp. 312–321). USA: ACM. https://doi.org/10.1145/1409635.1409677.

Zheng, Y., Liu, L., Wang, L., & Xie, X. (2008b). Learning transportation mode from raw Gps data for geographic applications on the web. In *Proceedings of the 17th International Conference on World Wide Web, WWW '08*, (pp. 247–256). New York, NY, USA: ACM. https://doi.org/10.1145/1367497.1367532.

Zheng, Y., Xie, X., & Ma, W.Y. (2010). Geolife: A collaborative social networking service among user, location and trajectory. *IEEE Data Eng. Bull., 33*(2), 32–39.

Chapter 9
Uncovering Geo-Social Semantics from the Twitter Mention Network: An Integrated Approach Using Spatial Network Smoothing and Topic Modeling

Caglar Koylu

9.1 Introduction

Advancements in mobile technology and wide use of online social networks have enabled large scale structural and geographic analysis of social ties and human communication. Previous studies utilized user generated textual communication data such as geo-referenced tweets and messages exchanged in online platforms and metadata from call detail records to study the effect of geographic proximity on social interactions (Backstrom et al. 2010; Han et al. 2017); the influence of information diffusion and social networks on real-world geographic events such as demonstrations, protests, and group activities (Vasi and Suh 2013); and structural and geographic characteristics of the communication network (Kylasa et al. 2015; Takhteyev et al. 2012). Although such studies use information flows to model social interactions, they often are *content agnostic*—ignore the content of the information exchanged between the individuals of the network (Hansen 1999). However, user generated content can indicate underlying interpersonal, ideological, structural and even geographic relationships between people (Lin et al. 2016).

User generated content have been the focus of researchers in information and communication sciences as well as computational linguistics. Despite the efforts that incorporate latent semantic analysis and probabilistic models to extract common topics and themes from large textual data, there has been little work (Chen et al. 2016; Kim et al. 2016) that focus on understanding of geo-social semantics of interpersonal communication, i.e., how the semantics of information vary based on the geographic locations and communication ties among individuals.

This paper introduces an approach to extracting and visualizing geo-social semantics from a geographically-embedded communication network. Different

C. Koylu (✉)
Department of Geographical and Sustainability Sciences,
University of Iowa, 316 Jessup Hall, Iowa City, IA, USA
e-mail: caglar-koylu@uiowa.edu

© Springer International Publishing AG, part of Springer Nature 2018
S.-L. Shaw and D. Sui (eds.), *Human Dynamics Research in Smart
and Connected Communities*, Human Dynamics in Smart Cities,
https://doi.org/10.1007/978-3-319-73247-3_9

from the previous work that examine the geographic variation in the content produced by individuals, this paper presents an analysis of reciprocal conversations among individuals using an integrated approach of spatial network smoothing and topic modeling. To demonstrate the approach, over 700 million geo-located tweets in the U.S. from Aug. 1, 2015 to Aug. 1, 2016 were analyzed. First, geo-located tweets were preprocessed to extract mention tweets between personal accounts. Second, a geo-located reciprocal mention network was constructed in which a node represents an individual and a link represents the collection of mentions and replies between two individuals. Within the geo-located reciprocal mention network, each individual was assigned to an areal boundary (i.e., county) for sustaining the privacy of the user and make use of place-tagged tweets. Messages among every pair of individuals were then combined into a collection of documents such as chat histories. Third, probabilistic topic modeling was performed on the collection of documents to classify each chat history into a multivariate set of topics with differing probabilities. Fourth, the individual-to-individual reciprocal mention network with classified conversations was summarized into an area-to-area network by smoothing the ego-centric network of reciprocal connections per area. Finally, topical probabilities were calculated and mapped for each area to reveal geographic and semantic patterns of communication.

9.2   Background and Related Work

In the following sub-sections, the use of Twitter as an interpersonal communication network is discussed, and a review of related work that examines the structural, geographic and semantic patterns of communication networks is provided.

9.2.1   Twitter as a Communication Network

Due to data availability and functional relationships between its users such as follow, reply, mention and retweet, Twitter has become one of the most studied communication networks. Follower, favorite and retweet functions are often used for broadcasting information and studying the process of information diffusion. On the other hand, the form of direct communication among individuals is either through private messages or replies and mentions. While a reply is a response to another user's tweet that begins with the @username of the person that she/he is replying to, a mention is a tweet that contains another user's @username anywhere in the body of the tweet. Mentions and replies allow users to join conversations on Twitter, which social interaction could be inferred as a means of direct personal communication. The context of communication can also be inferred by close observation of the textual content of the messages being exchanged.

Previous studies (Compton et al. 2014; Jurgens 2013; Yamaguchi et al. 2013) revealed that user mentions on Twitter occur between users that are in close geographic proximity. To understand the structural characteristics, Kato et al. (2012) compared the favorite, follow and mention networks and found that all three networks are scale-free in degree distribution; and they reveal similar predominant network motifs that highlight mutual links. Cogan et al. (2012) reconstructed evolving graphs of user mentions and replies on Twitter around a particular message content and found two common typologies. The first one is "path", which illustrate back-and-forth conversations in a group of connected users. The second typology is "star", which corresponds to conversations where a single user generates a tweet to which a large number of people reply, however, the users do not respond to each other's replies. This paper focuses on the "path" typology, to study reciprocal communication among the users through the use of user mentions on Twitter.

9.2.2 Topic Analysis

A variety of methodologies including wavelet analysis (Weng and Lee 2011), principal component analysis (PCA) (Kondor et al. 2013), support vector machine (SVM) (2010) and generative models (Eisenstein et al. 2011) have been used to provide automatic or semi-automatic detection of relevant themes from Twitter data. Computational and semantic analysis techniques have been developed to infer human behavior, ideological and attitudinal similarity between individuals (Adamic et al. 2014), common topics and way of speaking (McCallum et al. 2007), and group identities (Tamburrini et al. 2015). Moreover, semantic analysis and probabilistic models such as Latent Dirichlet Allocation (LDA) (Chae et al. 2012; Hu and Ester 2013; Hu et al. 2015; Liu et al. 2013; Pozdnoukhov and Kaiser 2011; Sakaki et al. 2010; Zhang et al. 2009) have been successfully employed to detect geographic events, recommend places, and friends based on user location, and similarity of shared content between users in social media posts.

LDA is based on term frequency-inverse document frequency (TF-IDF) (Salton and McGill 1983), which is a statistic that takes into account the frequency of words in the corpus and reflects how important each word is to a document in a collection of documents or corpus. The TF-IDF value increases proportionally to the number of times a word appears in a document. A tweet can contain up to 140 characters which do not allow multiple co-occurrences of words being used within the same tweet. Thus, training a topic model with short documents (i.e., individual tweets) results in unstable classifications with increased uncertainty due to the severe data sparsity (Yan et al. 2013). Several methods have been proposed to address the issue which suggest combining multiple tweets into document bins. Grant et al. (2011) aggregated tweets into buckets based on a group of similarity measures. Hong and Davison (2010) showed that training a topic model on aggregated tweets by users provide a higher quality and significantly better performance in classifying tweets.

In addition to aggregating tweets by similarity and user, Malik et al. (2013) combined tweets into bins that cover a given time range which allows the discovery of temporal changes in topics. Gerber (2014) employed space-time binning and compiled tweets into a single document based on a time window and a spatial neighborhood. Different from these studies, in this paper, tweets exchanged among a pair of individuals are combined into a document such as a chat history, and topic modeling is performed on the collection documents in order to discover themes of conversations.

9.3 Methodology

9.3.1 Data Cleaning

The Twitter Streaming API is used to collect geo-located tweets using a geographic bounding box. Tweets with exact geographic coordinates (geo-tagged tweets) and place names (place-tagged tweets) which correspond to an area (e.g., city, neighborhood) are used, while place-tagged tweets at state or country level are disregarded. Geo-located tweets are preprocessed to extract mention tweets between personal accounts. The metadata provided by the API is used to filter the tweets and users. Each tweet includes an attribute that contains whether the tweet was generated using an external application, and what that application was. A review of the contents produced by each of these applications is performed to filter tweets from non-personal user accounts such as TweetMyJobs, which is used to recruit employees, local weather reports, emergency reports, traffic crash reports, news feeds and etc. Also, tweets generated by a number of external applications (e.g., Foursquare and Instagram) are removed. Most of those tweets produced by external applications do not include conversational context. For example, Foursquare enables automatic generation of a tweet's content with a standard text to indicate a user's location: "I'm at Smyrna; TN in Smyrna; TN". In addition, tweets from users with more than 3000 followers are removed to prevent any bias caused by a large number of user mentions attracted by a few users, i.e., celebrities (Lansley and Longley 2016).

9.3.2 Identifying and Locating Pairs of Reciprocal Communication

One can construct an individual-to-individual communication network, where a node represents a user, and a link represents a tweet sent from user A to user B (whom user A mentions or replies to). Replies and mentions are embodied within the message of the tweet and can be downloaded using the Twitter's streaming API. A geo-located tweet includes only the location of the sender who mentions or replies to another user (recipient). A representative location of the recipient in a mention can be derived only if the recipient has at least one geo-located tweet in the sample. In this paper, tweets from users who mention and reply each other at least once, and whose locations are known are used.

Since individuals are mobile, locations of tweets from each user are variable across space. A geo-located reciprocal mention network is constructed in which a node represents an individual and a link represents the collection of mentions and replies between two individuals. Within the geo-located reciprocal mention network, each individual is assigned to an areal boundary (i.e., county) for sustaining the privacy of the user and make use of the place-tagged tweets that are in lower resolution (i.e., an areal boundary instead of exact coordinates). Tweet locations are overlayed with census data (e.g., county boundaries) to identify a home area (e.g., county) for each user based on the most frequent tweet location. Another commonly used strategy is to determine the home location based on tweets posted at night time where individuals are assumed to be home. In this paper, geo-located tweets with exact coordinates and place names that corresponded to an area at least at city scale are used.

9.3.3 Topic Modeling of Interpersonal Communication

Messages among every pair of individuals are combined into a collection of documents such as chat histories. To classify the content of each chat history, a probabilistic topic model, LDA is performed. LDA provides a model of documents that assumes a collection of k topics defined as a multinomial distribution over words. In this paper, a document corresponds to a chat history which contains all the mention and reply tweets exchanged between a pair of users. This strategy allows classifying conversations rather than tweets from a user, tweets from certain time periods, or tweets from certain locations.

$$P(Z|W,D) = \frac{W_{Z+\beta w}}{total\ tokens\ in\ Z + \beta} * D_{Z+\propto}$$

For each possible topic Z, $P(Z|W,D)$ is the probability that word W came from document D, which is calculated by the multiplication of $W_{Z+\beta w}$ (i.e., the frequency of W in Z), by $D_{Z+\propto}$ (i.e., the number of other words in document D that already belong to Z). β and β_w are hyper-parameters that represent the chance that word W belongs to topic Z even if it is nowhere else associated with Z (Blei et al. 2003). Based on this formula, LDA iteratively goes through the collection, word by word, and reassigns each word to a topic. Words become more common in topics where they have higher frequencies; and thus, topics become more common in documents where they occur more often. After each iteration, the model becomes more consistent as topics with specific words and documents. The model eventually reaches an equilibrium that is as consistent as the collection allows. However, it is not possible to obtain a perfectly consistent model because topics and words do not have a one-to-one relationship (Underwood 2012). Mallet toolkit (McCallum 2002) is used to implement the LDA model and include stop words (e.g., commonly used words such as "the", "of", "am") from 28 languages prior to training the model. The topic model classifies conversations among each pair of individuals (i.e., all of the tweets between two users) with a mixture of latent topics in differing probabilities. For example, the conversations between an individual i and j might be classified as 50% about sports, 20% about fashion, 10% about food, and 10% about the other topics.

9.3.4 Estimating Topical Probabilities Over Geographic Areas

Once the conversations among each pair of individuals are classified into a set of topics, one can calculate the average topical probabilities per unit area. For example, among 1000 reciprocal user pairs in Kings County NY, one can calculate the average probability of a topic such as football, by simply adding the probability of the topic per user pair, and dividing the sum by the total number of user pairs. However, because of the variable population density some counties (or areas) will have a small number of user pairs. Thus, sparse sampling of the reciprocal user pairs across small areas (i.e., the small area problem in spatially-embedded networks) result in spurious variations, where a single node or connection is often too small (with insufficient data) for deriving stable statistical measures. To address the problem, adaptive kernel smoothing can be applied to network data in order to compute and map graph measures both in space (Koylu and Guo 2013) and space-time (Koylu et al. 2014). An adaptive kernel allows expanding the search space to include reciprocal connections of the geographic neighbors when the initial search space is found to be insufficient. This paper utilizes an adaptive kernel smoothing approach to consider connections from nearby areas. The approach is explained in the following sub-sections.

9.3.5 Neighborhood Selection and Kernel Smoothing

Neighborhood selection is the process of determining the reciprocal connections of each area which we can define as the ego-centric network. The ego-centric network includes not only the user pairs that both users reside within the same area but also the pairs that one of the users is in the area while the other user resides in a different area are also included. A major disadvantage of an adaptive kernel approach is over-smoothing the characteristics of areas with sparse observations especially when there is an area with dense observations in close geographic proximity.

For example, when the ego-centric network of a rural area includes reciprocal connections from a nearby urban setting, the network measure or topical probabilities for the rural area will resemble and be dominated by that of the urban area. Also, it is likely that the content of conversations in a rural area will be different than the conversations in an urban setting.

In order to limit the influence of areas with higher density of connections on the areas with sparse connections, a similarity threshold based on connection density is used in the neighborhood selection process. The distribution of the number of user pairs for all areas is considered, and one standard deviation of gross flow per area is used as the similarity threshold. Alternatively, one can incorporate a measure of topological similarity such as one that considers the network structure (e.g., triads), or measures such as centrality and clustering coefficient. Neighborhood selection and the adaptive kernel smoothing algorithm are introduced below.

Description of the Neighborhood Selection and Smoothing Algorithm

Definitions:

A_i	The area i for calculating the network measure. $A_i \in A$ (the total set of n unit areas, i.e., counties).
t	Neighborhood size threshold based on gross volume of flows.
WF_i	The number of reciprocal pairs within A_i.
σ	A similarity threshold to evaluate whether to include or not include a geographic neighbor into the neighborhood for smoothing. The standard deviation of WF_i is used as the threshold.
$N(A_i, t)$	The t-size neighborhood of an area A_i, $N(A_i, t)$, $t > 0$, is defined as the smallest $KNN(A_i, K) = \{A_j \in A \text{ and } \sqrt{(WF_i - WF_j)2} < \sigma\}$ that has a total size $\sum S_q > t$.
$LF(A_i, t)$	The list of flows within, and in and out of the neighborhood of $N(A_i, t)$.
$B(A_i, t)$	The bandwidth of the t-Size Neighborhood of A_i, is the radius of the smallest circle centered on A_i that covers all areas in the $N(A_i, t)$.
K	Kernel function. Uniform function is used where all weights = 1 in the neighborhood.
$F(A_i, t)$	$\sum_f^{LF(A_i,t)} Vol.(f) * weight(f)$: The weighted total volume of flows within, and in and out of the neighborhood of $N(A_i, t)$. In a kernel function (other than uniform) the weight of a flow can be calculated by the distance from the centroid of the area to the mid-point of the flow.

Steps:

(1) Compute WF, the number of reciprocal pairs within each unit area and σ, the standard deviation of number of reciprocal pairs for all units.
(2) Construct a Sort-tile-recursive (STR) tree for finding k-nearest-neighbors.
(3) Determine the neighborhood

   i. FOR each area A_i:
   ii.   IF $WF_i < t$
   iii.   Sort the nearest neighbors of A
   iv.   FOR each neighbor j
   v.   IF $\sqrt{(WF_i - WF_j)\,2} < \sigma$
   vi.   Add j into $N(A_i, t)$:
   vii.   FOR each flow in F_j
   viii.   IF flow does not exist in $F(A_i, t)$:
   ix.   Calculate flow weight based on K
   x.   Add [flow * weight] into $F(A_i, t)$

Given a positive neighborhood size threshold t based on the number of recip-rocal pairs, a t-size neighborhood is derived for each area $A_i \in A$, which is the smallest k-nearest-neighbor neighborhood of A_i (including itself) that meets the size constraint.

9.3.6 Calculating Topical Probabilities Per Area

Given the neighborhood and the list of reciprocal pairs, $LF(A_i, t)$, the topical probabilities per area can be calculated by using the following formula:

$$P_z(A_i|\theta) = \frac{\sum_{F_{ij} \in LF(A_i,t), i \neq j}^{F(Ai,t)} f(i,j) * p_z(i,j)}{F(A_i,t)}$$

$P_z(i,j)$ is the probability of topic z in conversations among the users i and j, which at least one of them reside in the neighborhood of A_i. $LF(A_i, t)$: is the list of reciprocal pairs in $N(A_i, t)$ (i.e., the neighborhood of A), and f_A is the number of reciprocal pairs in the neighborhood $N(A_i, t)$. $P_Z(A_i|\theta)$ is the average probability of topic z given all the topical probabilities (θ) in $N(A_i, t)$.

9.4 Results

Table 9.1 illustrates the descriptive statistics of the geo-located tweets within the Contiguous U.S. from Aug. 1, 2015 to Aug. 1, 2016. After the data cleaning and processing, there were 2,675,130 reciprocal contacts (distinct pairs of users that exchanged tweets among each other) with 33,141,460 mention tweets exchanged between those contacts. Similar to the findings of the previous work, the amount of

Table 9.1 Tweet and user statistics

Total tweets	700,078,319	Users	6,570,305
Tweets with mentions	221,030,872	Users > 3000 followers	249,847
Geo-located user mentions	71,438,987 (32%)	Users with tweets in only one county	1,433,870
Tweets exchanged among reciprocal contacts	33,141,460 (46%)	Users mentioned another user at least once	4,719,197
Reciprocal contacts	2,675,130	Users with reciprocal contacts	1,539,396

communication greatly decreased by increasing geographic distance. While 50% of the geo-located reciprocal communication pairs were within the same county and 77% were within the same state.

9.4.1 Topics of Interpersonal Communication

To evaluate the influence of parameter selection in the results of the topic model, a set of topic models were trained using 20, 50, and 100 topics with 2000 iterations. The topical overlap among the models with different parameters were evaluated using cosine similarity. The model with 50 topics was selected based on an evaluation of overlapping topics within the model as well as the distinctness of the topics as compared to the models with 20 and 100 topics. Measures of probability (P), entropy (E) and corpus distance (CD) were used to interpret the topic modeling results. The probability of a topic represents the proportion of the corpus assigned to the topic, and calculated by the ratio of the number of word tokens assigned to the topic, to the sum of the token counts for all topics. The most interesting topics reside within the range of non-extreme values whereas extreme values indicate unreliable topics. A small probability indicates that a topic may not be reliable because we do not have enough observations to examine the topic's word distribution. On the other hand, a large probability indicates extremely frequent topic, which could be considered as a collection of corpus specific stop-words. Document entropy illustrates whether a topic is distributed evenly over conversations among many users (high entropy), or occur a lot in a smaller number of conversations (low entropy). Corpus distance measures how far a topic is from the overall distribution of words in the corpus. A greater corpus distance means the topic is more distinct; a smaller distance means that the topic is more similar to the corpus distribution.

Table 9.2 illustrates thirteen topics that were selected based on a probability range of 0.01 and 0.03 (the median probability of all topics). The table includes both the words and metrics of each topic. Words are ranked by their probability of occurrence from the highest to the lowest. One can infer the latent topic using the combination of the words that commonly co-occur. Some of the latent topics such as "friends & family" and "couples" do not contain words that can be used to infer

Table 9.2 Thirteen latent topics with words and diagnostics measures P: Probability, E: Entropy, CD: Corpus distance

Topic	Words	P	E	CD
Football	Game, team, year, win, play, football, season, qb, won, games, big, fans, teams, beat, years, week, nfl, guy, coach, defense	0.035	10.66	1.48
Civil rights	Black, white, point, agree, women, isn, read, law, ppl, wrong, police, guns, problem, kids, racist, country, understand, true, cops, matter	0.030	10.54	1.65
Friends and family	Literally, bc, cute, tho, wow guys, wtf, true, ily, crying, wait, rn, idk, tweet, mom, thought, funny, ugh, honestly, bye	0.025	12.39	0.91
Couples	Baby, babe, beautiful, cute, wait, birthday, amazing, bae, sweet, girlfriend, perfect, heart, boyfriend, wcw, lucky, boo, months, princess, blessed, gorgeous	0.023	11.42	1.76
Weather	Snow, rain, weather, nice, cold, live, beach, storm, water, north, winter, south, week, long, beautiful, west, fun, year, weekend, hope	0.021	9.10	1.83
Faith	Sis, church, jesus, twug, amen, pastor, lord, bless, faith, blessed, christ, worship, plz, family, pray, twugs, cuffmedanny, praying, sunday, word	0.021	8.27	3.22
NBA	Team, game, year, lebron, win, play, curry, player, cavs, warriors, nba, kobe, won, steph, season, kd, games, finals, tho, ball	0.021	10.12	1.70
College sports	Congrats, team, luck, coach, game, win, boys, season, year, big, work, job, congratulations, girls, school, awesome, ready, support, week, 2016	0.019	10.58	2.05
Baseball and hockey	Game, team, year, win, baseball, season, games, play, cubs, fans, mets, guy, hockey, guys, series, won, years trade, teams, big	0.018	10.66	1.32
Learning	Team, students, work, awesome, join, excited, meeting, learning, support, event, amazing, community, check, sharing, congrats, ready, thx, job, fun, share	0.018	10.47	2.12
Primaries	Trump, vote, Hillary, Bernie, Cruz, Obama, Gop, president, party, Clinton, voting, Sanders, won, win, support, America, country, Donald, candidate, agree	0.016	10.17	1.96
Driving	Work, drive, money, pay, buy, ride, bike, lot, driving, nice, truck, parking, city, cars, house, park, bus, gas, street, live	0.012	11.36	1.49
Drinking	Beer, drinking, cheers, photo, wine, drink, coffee, nice, ipa, bar, beers, food, awesome, dinner, enjoy, fun, tap, delicious, lunch, bottle	0.010	9.77	2.64

the context of the conversation. These topics are formed by language elements used in conversational context including social media acronyms (e.g., bc, ppl, ily, idk, and etc.), or words in particular dialects (e.g., yall, bruh, ima, finna, and etc.). Latent topics of "football" and "civil rights" are among the most common topics. Although the data was captured during the primary elections, mentions about primary elections and candidates is among the low probability topics. Main reason for having a low probability distribution for political topics may be that the majority of political or election related conversations are likely to be among users who do not share geographic locations of their tweets. We can also attribute the lower probability for election related mentions to the fact that most election related content are produced and retweeted within highly segregated partisan networks where there are limited connections and conversations among left and right leaning users (Conover et al. 2011; Grabowicz et al. 2012). On the other hand, user mentions in a political context often occur within a single heterogeneous cluster of users in which opposite views interact with a much higher rate than in retweet networks. However, these clusters have been observed to be less dense than more homogenous clusters of retweets (Conover et al. 2011).

Latent topics derived by the topic model are often vague in terms of the sentiment and context of conversations. This is due to the loss of sentiment and context as a result of the bag of words approach used in topic modeling. For example, words such as won, vote, voting, win, support and agree are used with any of the candidates, however, the context for their usage is lost.

Table 9.3 represents outlier topics with high and low probabilities. "Birthday" topic has the highest probability among all topics, and represent happy birthday messages and celebrations. "Food" and "Fashion" related conversations are also quite common among the users. On the other hand, low probability topics indicate rare mentions. The two topics with the lowest probability represent mentions in languages other than English, i.e., Arabic and Spanish.

Table 9.3 Example outlier topics (high and low probability)

High probability topics				
Topic	Words	P	E	CD
Birthday	Birthday, hope, pretty, bday, beautiful, amazing, hbd, ily, gorgeous, awesome, enjoy, lots, babe, sweet, aw, wonderful, fun, wait, congrats, guys	0.111 (the highest prob.)	13.04	2.37
Food	Food, eat, chicken, cheese, pizza, eating, dinner, lunch, taco, breakfast, fries, sauce, hot, bacon, tacos, meat, burger, cook, wings, bread	0.061	11.19	2.50
Fashion	Hair, wear, black, wearing, color, white, cute, shirt, buy, red, cut, shoes, dress, makeup, blue, pretty, nice, long, tho, pink	0.037	11.64	1.79
Low probability topics				
Spanish	Gracias, ko, feliz, ang, batb, ng, mo, hola, amiga, saludos, dias, jajaja, nga, ay, ba, naman, quiero, noches, jajajaja, yan	0.004	7.54	2.50
Arabic	الي, لك, الله, بس, انا, والله, انت, مو, شي, اللي, بو, ي 笑, علي, ك,و, انت, لو, خير, يابو, تو, شاء	0.0006 (the lowest prob.)	7.26	5.76

9.4.2   Geo-Social Semantics of Interpersonal Communication

Using the adaptive kernel approach, one can produce a probability map for each topic for understanding the geo-social semantics of reciprocal mentions among users. Figure 9.1 illustrates the geographic distribution of two political topics: mentions of primary elections, and civil rights. Both maps in Fig. 9.1 has the same legend which allows comparing the resulting probabilities of the two topics. Probability value refers to the commonality of the topic mentioned among individuals for the ego-centric reciprocal network of each area on the map.

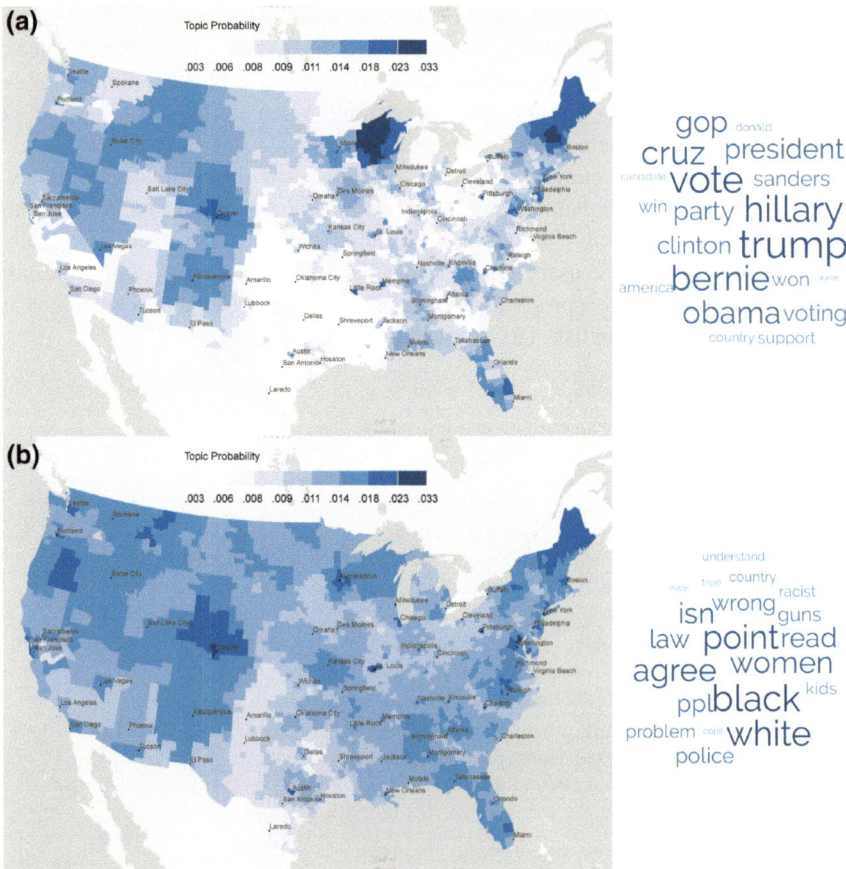

Fig. 9.1 Topic probabilities. **a** Primary candidates and elections. **b** Civil rights. While the topic of primary candidates and elections was prominent at locations of primary elections and core supporters of candidates; civil rights was a prominent topic across the whole country

The topic of primary elections consisted of candidate names, words of political context such as gop, obama, party, and candidate; and other election specific words such as vote, support, and win. On the other hand, civil rights topic was formed by commonly used words such as black, white, women, law, ppl (people), kids, police and guns, and words used in debates such as point, agree, isn (is not), read, wrong, problem, racist, country, understand, true and matter (Table 9.2). The word cloud represents frequently co-occurring words for each topic. Some words co-occur with a much higher frequency than others, which makes word clouds difficult to interpret. For example, the most commonly used word within the topic of primary elections was Trump, which occurred approximately three times the words Hillary and Bernie, and ten times the least frequent word agree. In order to make the word cloud more readable, font sizes are assigned based on the ranking of words within a topic. The larger the font size the highest the ranking of the word, which is assessed by its frequency within the topic.

From Fig. 9.1a we can infer that individuals were highly engaged in election related conversations in the North-East states of Vermont, New Hampshire and Maine; and in the north of Wisconsin and Michigan. Election related content was discussed within certain geographic locations that reflect the locations of primary elections and supporters of candidates. On the other hand, Fig. 9.1b highlights the metropolitan areas such as Denver, St. Louis, Washington D.C., Seattle, Portland, Minneapolis and New York City as hot spots of civil rights discussions. While the topic of primary candidates and elections highlighted localized clusters of high values in some metropolitan areas, and the North-East and rural areas in the north of Wisconsin; civil rights was a prominent topic across the whole country.

Figure 9.2 highlights a clustering of "faith" topic in the South, which peaked around the states of Tennessee, South Carolina and North Carolina. Although faith is a rare topic mentioned among individuals, the clustering of high topical probabilities align well with the religious regions of the US. It is striking that coastal areas do not have as high values as the inland areas in the South. There are also regional clusters of Idaho and the north of Nevada, and New Mexico. Besides these clusters there are also spikes of metropolitan suburbs with elevated probabilities. While the words that form this topic are coherent and mostly have religious context, there is an exception of the word "cuffmedanny" which is a hashtag used in a TV series. Presence of words about this TV show in mention tweets suggest second screening (Doughty et al. 2012), which refer to individuals that live-tweet during a broadcast. In this topic, the religious references and the show co-occurred in substantial portion of conversations.

Figure 9.3 illustrates the geographic distribution of the topic "NBA finals". Unsurprisingly, NBA finals were predominantly discussed in metropolitan areas with major NBA teams such as Cleveland, San Francisco and Oklahoma City. Similar to the candidate names in primary elections, this topic was formed by the names of NBA teams such as Cavs and Warriors, and NBA players such as Stephen Curry, LeBron James and Kevin Durant (KD).

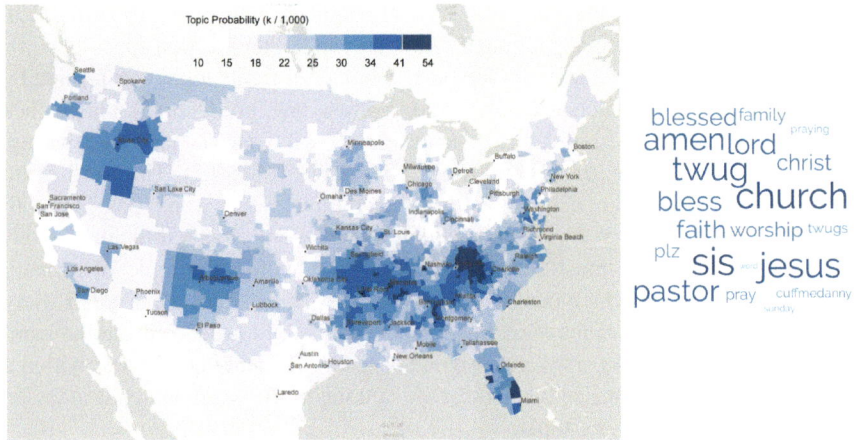

Fig. 9.2 Topical probabilities "Faith". Although faith is a rare topic mentioned among individuals, the clustering of high topical probabilities align well with the religious regions of the US

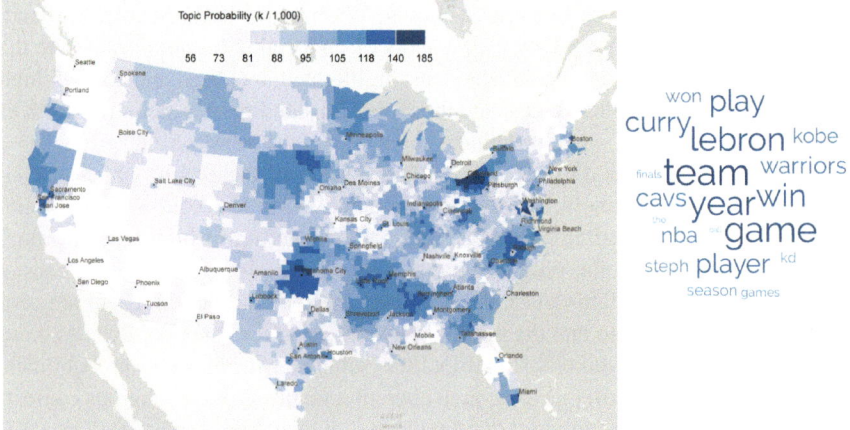

Fig. 9.3 Topic probabilities "NBA Finals". Unsurprisingly, NBA finals were predominantly discussed in metropolitan areas with major NBA teams such as Cleveland, San Francisco and Oklahoma City

9.5 Discussion and Conclusion

A novel approach for extracting topical themes and their spatial patterns from a geographically-embedded interpersonal communication network was presented. The approach was demonstrated using a year of geo-located reciprocal user mentions on Twitter. The results revealed varying geographic patterns of communication on topics such as civil rights, primary elections and candidates, sports, weather, faith, food and fashion. Extracted topics reflect geo-social dynamics of the society; way of speaking in the context of friendship, and couples; and linguistic variation and the use of social media acronyms. Unlike the given time period of the dataset which covers the entire period of primary elections, mentions about the primary candidates and elections were among the least prominent topics. On the other hand, mentions about civil rights, which include race, gender and gun rights were found to be among the highest probability topics, and widely discussed across the whole country. Although the tweets were collected during primary and presidential elections, political topics discovered from the reciprocal mentions focused more on civil rights rather than the candidates and primaries. Also, individuals were highly engaged in civil rights conversations across the country, whereas election related content was discussed within certain geographic locations that reflect the locations of primary elections and supporters of candidates.

There are a number of directions for future work. First, a major limitation of this study is that the temporal variation of the topics was ignored. The topic model can be trained to extract temporally varying topics and the evolution of topical content over time. Unsurprisingly, the topics extracted from reciprocal mentions align well with regional geographies of semantic content such as politics, faith and NBA. There is a need to compare the patterns of topics derived from the reciprocal communication of users with the content of the tweets generated, or retweeted without mentioning others. Such analysis would help us understand the semantic variation and the differences in geographic patterns between the interpersonal communication and user behavior for information broadcasting on Twitter.

References

Adamic, L. A., Lento, T. M., Adar, E., & Ng, P. C. (2014). Information evolution in social networks. *arXiv preprint* arXiv:1402.6792.

Backstrom, L., Sun, E., & Marlow, C. (2010, April 26–30). *Find me if you can: Improving geographical prediction with social and spatial proximity.* Paper Presented at the Proceedings of the 19th International Conference on World Wide Web, Raleigh, NC, USA.

Blei, D. M., Ng, A. Y., & Jordan, M. I. (2003). Latent dirichlet allocation. *The Journal of Machine Learning Research, 3,* 993–1022.

Chae, J., Thom, D., Bosch, H., Jang, Y., Maciejewski, R., Ebert, D. S., et al. (2012). *Spatiotemporal social media analytics for abnormal event detection and examination using seasonal-trend decomposition.* Paper Presented at the 2012 IEEE Conference on Visual Analytics Science and Technology (VAST).

Chen, S., Yuan, X., Wang, Z., Guo, C., Liang, J., Wang, Z., et al. (2016). Interactive visual discovering of movement patterns from sparsely sampled geo-tagged social media data. *IEEE Transactions on Visualization and Computer Graphics, 22*(1), 270–279.

Cogan, P., Andrews, M., Bradonjic, M., Kennedy, W. S., Sala, A., & Tucci, G. (2012). *Reconstruction and analysis of Twitter conversation graphs.* Paper Presented at the Proceedings of the First ACM International Workshop on Hot Topics on Interdisciplinary Social Networks Research.

Compton, R., Jurgens, D., & Allen, D. (2014). *Geotagging one hundred million Twitter accounts with total variation minimization.* Paper Presented at the 2014 IEEE International Conference on Big Data (Big Data).

Conover, M., Ratkiewicz, J., Francisco, M., Gonçalves, B., Menczer, F., & Flammini, A. (2011). *Political Polarization on Twitter.* Paper presented at the ICWSM.

Doughty, M., Rowland, D., & Lawson, S. (2012). *Who is on your sofa? TV audience communities and second screening social networks.* Paper Presented at the Proceedings of the 10th European Conference on Interactive TV and Video.

Eisenstein, J., Ahmed, A., & Xing, E. P. (2011). *Sparse additive generative models of text.* Paper Presented at the Proceedings of the 28th International Conference on Machine Learning (ICML-11).

Gerber, M. S. (2014). Predicting crime using Twitter and kernel density estimation. *Decision Support Systems, 61,* 115–125.

Grabowicz, P. A., Ramasco, J. J., Moro, E., Pujol, J. M., & Eguiluz, V. M. (2012). Social features of online networks: The strength of intermediary ties in online social media. *PLoS ONE, 7*(1), e29358.

Grant, C. E., George, C. P., Jenneisch, C., & Wilson, J. N. (2011). *Online topic modeling for real-time Twitter search.* Paper Presented at the TREC.

Hansen, M. T. (1999). The search-transfer problem: The role of weak ties in sharing knowledge across organization subunits. *Administrative Science Quarterly, 44*(1), 82–111. https://doi.org/10.2307/2667032.

Han, S. Y., Tsou, M.-H., & Clarke, K. C. (2017). Revisiting the death of geography in the era of Big Data: The friction of distance in cyberspace and real space. *International Journal of Digital Earth,* 1–19.

Hong, L., & Davison, B. D. (2010). *Empirical study of topic modeling in Twitter.* Paper Presented at the Proceedings of the First Workshop on Social Media Analytics.

Hu, B., & Ester, M. (2013). *Spatial topic modeling in online social media for location recommendation.* Paper Presented at the Proceedings of the 7th ACM Conference on Recommender Systems.

Hu, Y. J., Gao, S., Janowicz, K., Yu, B. L., Li, W. W., & Prasad, S. (2015). Extracting and understanding urban areas of interest using geotagged photos. *Computers, Environment and Urban Systems, 54,* 240–254. https://doi.org/10.1016/j.compenvurbsys.2015.09.001.

Jurgens, D. (2013). That's what friends are for: Inferring location in online social media platforms based on social relationships. *ICWSM, 13,* 273–282.

Kato, S., Koide, A., Fushimi, T., Saito, K., & Motoda, H. (2012). *Network analysis of three Twitter functions: Favorite, follow and mention.* Paper Presented at the Pacific Rim Knowledge Acquisition Workshop.

Kim, K. S., Kojima, I., & Ogawa, H. (2016). Discovery of local topics by using latent spatio-temporal relationships in geo-social media. *International Journal of Geographical Information Science, 30*(9), 1899–1922. https://doi.org/10.1080/13658816.2016.1146956.

Kondor, D., Csabai, I., Dobos, L., Szule, J., Barankai, N., Hanyecz, T., ... Vattey, G. (2013). *Using Robust PCA to estimate regional characteristics of language use from geo-tagged Twitter messages.* Paper Presented at the 2013 IEEE 4th International Conference on Cognitive Infocommunications (CogInfoCom).

Koylu, C., & Guo, D. (2013). Smoothing locational measures in spatial interaction networks. *Computers, Environment and Urban Systems, 41,* 12–25. https://doi.org/10.1016/j.compenvurbsys.2013.03.001.

Koylu, C., Guo, D., Kasakoff, A., & Adams, J. W. (2014). Mapping family connectedness across space and time. *Cartography and Geographic Information Science, 41*(1), 14–26.

Kylasa, S. B., Kollias, G., & Grama, A. (2015). *Social ties and checkin sites: Connections and latent structures in location based social networks.* Paper Presented at the Proceedings of the 2015 IEEE/ACM International Conference on Advances in Social Networks Analysis and Mining 2015.

Lansley, G., & Longley, P. A. (2016). The geography of Twitter topics in London. *Computers, Environment and Urban Systems, 58,* 85–96.

Lin, Y.-R., Margolin, D., & Lazer, D. (2016). Uncovering social semantics from textual traces: A theory-driven approach and evidence from public statements of U.S. Members of Congress.

Journal of the Association for Information Science and Technology, 67, 2072–2089. https://doi.org/10.1002/asi.23540.

Liu, X., Tang, K., Hancock, J., Han, J., Song, M., Xu, R., et al. (2013). A text cube approach to human, social and cultural behavior in the twitter stream. *Social computing, behavioral-cultural modeling and prediction* (pp. 321–330). Berlin: Springer.

Malik, S., Smith, A., Hawes, T., Papadatos, P., Li, J., Dunne, C., & Shneiderman, B. (2013). *TopicFlow: Visualizing topic alignment of Twitter data over time.* Paper Presented at the Proceedings of the 2013 IEEE/ACM International Conference on Advances in Social Networks Analysis and Mining, Niagara, Ontario, Canada.

McCallum, A. K. (2002). MALLET: A machine learning for language toolkit. http://mallet.cs.umass.edu.

McCallum, A., Wang, X., & Corrada-Emmanuel, A. (2007). Topic and role discovery in social networks with experiments on enron and academic email. *Journal of Artificial Intelligence Research, 30,* 249–272.

Pozdnoukhov, A., & Kaiser, C. (2011). *Space-time dynamics of topics in streaming text.* Paper Presented at the Proceedings of the 3rd ACM SIGSPATIAL International Workshop on Location-Based Social Networks.

Sakaki, T., Okazaki, M., & Matsuo, Y. (2010). *Earthquake shakes Twitter users: Real-time event detection by social sensors.* Paper Presented at the Proceedings of the 19th International Conference on World Wide Web.

Salton, G., & McGill, M. J. (1983). Introduction to Modern Information Retrieval.

Takhteyev, Y., Gruzd, A., & Wellman, B. (2012). Geography of Twitter networks. *Social Networks, 34*(1), 73–81. https://doi.org/10.1016/j.socnet.2011.05.006.

Tamburrini, N., Cinnirella, M., Jansen, V. A., & Bryden, J. (2015). Twitter users change word usage according to conversation-partner social identity. *Social Networks, 40,* 84–89.

Underwood, T. (2012). Topic modeling made just simple enough. Retrieved from http://tedunderwood.com/2012/04/07/topic-modeling-made-just-simple-enough/.

Vasi, I. B., & Suh, C. S. (2013). *Protest in the Internet Age: Public attention, social media, and the spread of "Occupy" protests in the United States.*

Weng, J., & Lee, B.-S. (2011). Event detection in Twitter. *ICWSM, 11,* 401–408.

Yamaguchi, Y., Amagasa, T., & Kitagawa, H. (2013). *Landmark-based user location inference in social media.* Paper Presented at the Proceedings of the first ACM Conference on Online Social Networks.

Yan, X., Guo, J., Lan, Y., & Cheng, X. (2013). *A biterm topic model for short texts.* Paper Presented at the Proceedings of the 22nd International Conference on World Wide Web.

Zhang, D., Zhai, C., & Han, J. (2009). *Topic cube: Topic modeling for OLAP on multidimensional text databases.* Paper Presented at the SDM.

Author Biography

Caglar Koylu (Ph.D., University of South Carolina), Assistant Professor of GIScience at the department of Geographical and Sustainability Sciences at the University of Iowa. Dr. Koylu's research interests lie at the intersection of spatial data mining, big data analytics, network science, machine learning, geovisual analytics and human-computer interaction. His special research interests in human dynamics research are developing new theories, visual and computational approaches to understanding complex patterns from large geo-social networks, i.e., networks embedded in geographic space and time such as migration, human mobility and communication, flows of information, innovation and capital, social and family networks and commodity flows.

Chapter 10
Grouping People in Cities: From Space-Time to Place-Time Based Profiling

Tao Cheng and Jianan Shen

10.1 Introduction

Before the Internet and mobile devices become part of our everyday life, early studies of human activity patterns were confined to traditional statistical and survey studies that involved tracking, logging, managing and analysing the massive and detailed life cycles of individuals. Modern ubiquitous telecommunication and sensor technologies make this process simpler. Large scale data collection concerning the movement trajectories of massive numbers of users, such as GPS data and mobile phone user data, has become technically feasible and economically affordable. The information contained in the moving trajectory datasets makes them particularly suitable for the research of human dynamics (Tsou 2015).

Location-based service (LBS) has been an especially popular industry with the above-mentioned technologies in recent years. Some of the LBS applications, such as Foursquare and Twitter, have penetrated all aspects of daily life and provided a huge amount of data, recording the "check-ins" and place visiting behaviours of millions of users (Shaw et al. 2016). These data provide "3W" information, namely "when s/he visited the place, where the place is located, and what the place/activity is about", thus capturing semantically meaningful snapshots of personal behaviour patterns which enables the analysis "who specific people are". Shen and Cheng (2016) developed a framework to uncover space-time activity patterns from individual's movement trajectory data and segregate users into subgroups according to these patterns. They introduced the concept of '*where, when and how long you stay*

T. Cheng (✉) · J. Shen
SpaceTimeLab for Big Data Analytics, Department of Civil,
Environmental & Geomatic Engineering, University College London,
Gower Street, London WC1E 6BT, UK
e-mail: tao.cheng@ucl.ac.uk

J. Shen
e-mail: jianan.shen.13@ucl.ac.uk

is who you are', which is especially suited for analysing human dynamics of relatively small populations in small areas.

Goodchild (2015) proposed that "Platial views offer new insights beyond traditional spatial perspectives as human activity is more aligned with place rather than geometric space", which indicated the importance of "what the place/activity is about" over the physical location (where). Here, we further develop the concept of group human dynamics into '*what place, when and how long you stay is who you are*', expanding the focus from *location* to *place* by integrating the semantic meaning of places into the analysis. This evolution also enables us to analyse a large population with much higher heterogeneity and dynamism in a large city-scale area.

To take advantage of spatial, temporal and semantic information, we attempt to establish a universal framework that enables comprehensive analysis of space, time and activity in order to study how people spend their days differently. The framework segregates individuals into subgroups based upon what (*place*), when (*time*) and how long (*duration*) certain activities take place for each individual. We enrich the semantics and significance of places by using Points of Interest (POIs) in a city. An individual's profile is described as a summary of his/her time budget in different places, which are further clustered into groups of people with similar space-time activity patterns. This is tested by grouping the foot patrol police officers in London based on their GPS trajectories. This specific application demonstrates the potential of using the framework and methods for other GPS-based mobility datasets.

10.2   Related Works

We first discuss the related works in detecting regions of interest (ROIs) based upon GPS data, then brief on semantic enrichment of the ROIs. At last, methods of clustering semantic trajectories and behaviour patterns are summarised.

10.2.1   Detecting Regions of Interests

A generic and well-known paradigm of pre-processing the raw GPS trajectories is trip segregation, which separates the GPS points of every individual object into several consecutive trips. Every trip can be further transformed and divided into 'move episodes' and 'stop episodes' (Alvares et al. 2007; Spaccapietra et al. 2008), and semantic meaning or contextual information can be added to each episode.

The semantic analysis mostly focuses on the stay points of the stop episodes in which the moving objects stop moving, move in a small and confined area or move at a speed lower than a predefined speed threshold (Parent et al. 2013; Ying et al. 2013; Zheng et al. 2009). Areas with high density aggregation of identified stay points are considered as interesting regions or regions of interests (ROI) for the moving objects. For detecting ROIs, Zhao (2011) used minimum bounding boxes (MBB) to define the highly active region of moving objects' trajectories. Yan et al. (2013) used spatial bounding rectangle of the stop episode or the centre of the stay points to find interesting places.

Density based clustering (DBSCAN) is the most common methods for ROI detection as well (Giannotti et al. 2007; Güting et al. 2006; Parent et al. 2013; Li et al. 2010; Karlis and Saygin 2009; Palma et al. 2009). Researchers also improved and adapted the traditional density based method to cater for different case studies and research purposes. Zimmermann et al. (2009) designed a time-based OPTICS algorithm to cluster stay points, considering both spatial and temporal properties of a trajectory. Similarly, Shen and Cheng (2016) applied another variant of DBSCAN, ST-DBSCAN (Birant and Kut 2007), to detect spatio-temporal regions of interests (ST-ROI) and generated interesting regions with not only spatial location but also time span.

10.2.2 Enriching the Semantic Meaning of ROIs with POI Information

A semantic trajectory is a series of moving and stopping behaviours whose predicate bears on contextual information. POI and land use data, which was originally used to answer user-centred navigation questions such as 'What is here?', can also be applied to enrich location data with behavioural meaning (Krüger et al. 2015). Hence, information of the context objects that fall in the adjacency of the ROIs are analysed for semantic enrichment. For examples, Yan et al. (2013) used well defined land use and POI data to annotate the meaning of places and road segments. Krueger et al. (2013, 2015) also use POI for the same purpose and made a simple assumption that the meaning of a place is determined by the dominant type of POIs in the region.

Furthermore, Krüger et al. (2013) considered distance between the POIs and the location of individuals' activities to improve the semantic enrichment and compared the suitability of the POI data provided by Foursquare, Facebook and Google for semantic analysis. In their work, POIs closer to the individual's locations are considered to impose bigger impact on the individual than other POIs in the area. Damiani et al. (2011) also weighted POIs differently for protection of the individuals' privacy in sensitive stops. Since POIs nearby the individuals' locations may have a different degree of sensitivity according to their semantic meanings, lowering the significance of certain types of POIs can keep the individual sensitive behaviours from being compromised. For instance, stopping at normal restaurants are considered less sensitive than being in hospitals, and restaurants and hospitals should be given different weights. One noticeable fact is that all the land use and POI data used for semantic analysis in the works above are hierarchically classified into multiple categories and subcategories so that the sematic meaning can be summarised in different levels of details.

Besides the spatial context, temporal information is also used for semantic enrichment since most trajectories contain temporal records. Liao et al. (2006) proposed that different activities have different temporal durations and temporal patterns, which can be used to distinguish activities nearby multiple POIs. Andrienko et al. (2013) suggested to interpret semantic meanings of places based on cyclic temporal patterns of visiting times. Reumers et al. (2013) designed a classification tree to identify semantic places, which purely relies on temporal stop durations.

10.2.3 Clustering Analysis of Semantic Trajectories

Most commonly, the behaviour similarity computation is based on a distance or dissimilarity measure. The usual distance measures are spatial distance and temporal distance (Lee et al. 2007). Zheng et al. (2009) defined a similarity metric to find people of similar behaviour patterns based on the number of commonly visited POIs. Shen and Cheng (2016) defined the dissimilarity of behaviours base on the Jessen Shannon Distance (JSD) between the time budget allocation profiles (i.e. space time profiles) of individual persons. Yan et al. (2013) also compared the stopping and moving time distribution of vehicle trajectories in different types of semantic places to show the behavioural differences of cars, buses, taxis, and trucks.

10.3 Methodology

In our previous work (Shen and Cheng 2016), the concept of '*where, when and how long you stay is who you are*' is realised via five steps:

(1) **Space-Time regions of interest (ST-ROI) detection**: Discovering interesting places in space and time, determining where the places are and when they are 'interesting';
(2) **ST-profiling**: Describing individual profiles with the users' time allocation and staying behaviour in different places and time periods;
(3) **Profiling comparison**: Defining similarity metrics for measuring and quantify the behavioural differences between individual profiles;
(4) **Hierarchical clustering**: Grouping people with similar behavioural patterns based on the defined similarity metrics;
(5) **Semantic validation**: Explaining how these groups are formed.

This framework is specifically suited for analysing human dynamics of relatively small populations in small areas. To extend this concept to a large population that will generate many ST-ROIs, the comparison of profiling would be time consuming. To overcome this limit, a new concept of '*what place, when and how long you stay is who you are*' is proposed here. The main difference between this newly-proposed concept and the original is that the focus has been switched from "*Where*" (the geographical location) to "*What place*" (the semantic meaning of the place). The reason for this is that in a big city, there are many places offering similar functions and services, such as gyms in different districts of a city. Individuals doing the same activity at different places, rather than people doing different things in neighbouring areas, should be considered similar in terms of behavioural patterns.

To make sense of these ST-ROIs, we can use POIs feature type data to extract the semantic meaning of the ST-ROIs and summarise them into generic types, such as tube stations, shops, public infrastructures and museums. In this way, the profiles of individuals will be made up of the time allocated to each generic ST-ROI type, not each specific ST-ROI that is associated with a spatial location. This means that the concept can evolve from '*where (ST-ROIs), when and how long you stay is who you are*' to '*what place (ST-ROI types), when and how long you stay is who you are*', so that the precise locations of places are replaced by their meanings and people undertaking similar things in difference locations can be grouped together.

This new concept can be realised via four steps based upon movement trajectory data, as follows:

(1) **ST-ROI detection**: Extracting ST-ROIs for individuals. This step is identical to the first step in previous work (Shen and Cheng 2016), but is applied to a much larger dataset in a much larger study area;
(2) **Semantic enrichment of ST-ROIs**: Using and weighting POI data to analyse the semantic meaning of extracted ST-ROIs;
(3) **Building user semantic profiles**: Simplifying users' activities as one-dimensional time budget allocations to different semantic ST-ROIs to build a semantic profile for each user;
(4) **Hierarchical clustering**: Grouping similar semantic profiles of users to sum-marise representative behavioural patterns. No further semantic explanation like the final step in the previous work (Shen and Cheng 2016) is necessary, because the semantic analysis of places is already incorporated into the semantic enrichment process.

Step 1 here is exactly the same as Step 1 in our previous work (Shen and Cheng 2016). ST-DBSCAN (Birant and Kut 2007) was chosen to detect places and time periods with high visit intensity because of its capability of clustering objects with a combination of both spatial and temporal measurements, and detecting noise when different densities exist. The next two sections will explain Steps 2, 3 and 4 in greater detail.

We first applied this new method on the same trajectory dataset used in our previous work (Shen and Cheng 2016) to show the differences the new method can make on in the same study area. The dataset was generated by foot patrol police officers' GPS-integrated portable radio sets, which record positions every 10 min. In the chosen study period of February 2012, 355 officers generated 84,027 point records with call sign information, device IDs and time-stamped geo-locations. For further details, please refer to Shen and Cheng (2016). Afterwards, the proposed method was tested with an extended and multiple-borough case study in Sect. 10.4.

10.3.1 Semantic Enrichment of ST-ROIs

Enriching the semantic meaning of the ST-ROI can help better answer "what the place/activity is about" and explain why individuals allocate their time differently among ST-ROIs. This is a crucial step towards evolving the concept of '*where (ST-ROIs), when and how long you stay is who you are*' from our previous work to '*what place (ST-ROIs types), when and how long you stay is who you are*' that is introduced in the present article.

The semantic meanings of the ST-ROIs are achieved through the following steps. First, the spatial boundaries of the ST-ROIs are generated by creating a 20-m buffer zone for the minimal convex hull covering all the stay points in a ST-ROI, then the buildings and POIs that locate in the expanded convex hull area (i.e. the buffer zone of the minimum convex hull) are extracted. Lastly, the various categories of POIs are weighted using term frequency–inverse document frequency (TF–IDF) to move their semantic contribution index to the ST-ROI area into which they fall.

10.3.2 Bounding Convex Hulls of ST-ROIs

Figure 10.1 shows 28 ST-ROIs detected in police patrol activities in the Borough of Camden, London, based upon an Automatic Personnel Location System (APLS) dataset provided by the London metropolitan police. As can be seen in Fig. 10.1a, the visualisation is a 3D dot distribution map. Dot distribution maps represent spatial distribution of geo-referenced data using points as a basic graphical element (Slocum et al. 2008). Every point represents a datum with geo-location information. The points can only show the density of an area, rather than depict the clear boundary and exact location of the area. We must therefore implement a method that allows us to define the boundary of the ST-ROI sets before explaining the semantic meaning within the area. Convex hulls are usually used to turn point-based objects into spatial areas, and the minimal convex hull bounding method can create a polygon area enclosing all the stay points of each ST-ROI, making it the most ideal way to serve this purpose (Andrew 1979). In this step, we used the parallel spatial retrieving method (Miller and Stout 1988) to find the convex hulls that define the spatial boundaries of the ST-ROIs (see Fig. 10.1b). In practice, a 20 m buffer zone (i.e. the expanded convex hull) is used to define the actual coverage area of each ST-ROI. This buffer distance is set with the consideration that all the recorded GPS points have spatial errors and 20 m is the mean value of error in our dataset.

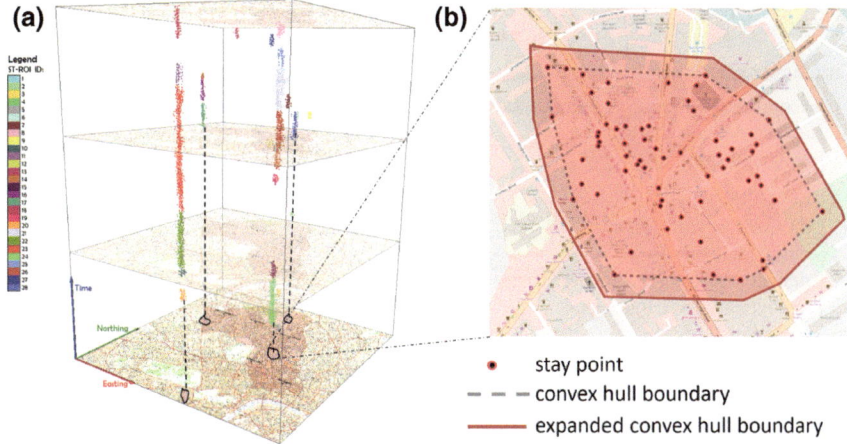

Fig. 10.1 a The 28 ST-ROIs found via the ST-DBSCAN method in our previous prototype framework (Shen and Cheng 2016); **b** Bounding convex hull and its 20 m buffer zone enclosing the stay points are generated to define the spatial area covered by the ST-ROIs

10.3.3   Semantic Meaning of ST-ROIs

A POI dataset contains the information of all the public buildings (POIs) that can be summarised to interpret the semantic meaning of a place in which they exist (Alvares et al. 2007; Alves et al. 2011; Braun et al. 2010). In order to understand the staying behaviour within each ST-ROI area, we used POI data to depict the functional images of the ST-ROIs and enrich the semantic meaning of users' visits to these ST-ROIs. The POI dataset used in the case study contains the information of a wide range of finely categorised infrastructures and buildings that offer different services and utilities (Ordnance Survey 2016). Similar to the hierarchical category structure of POI information used by Krüger et al. (2015) and Yan et al. (2013), the official Ordnance Survey POI classification scheme has three levels, with nine major categories and 52 sub-categories that can be further broken down into more than 600 detailed classes. Customers can adopt the official classification scheme defined by the Ordnance Survey. They are also encouraged to customise their own classification by selecting POIs from any combination of categories or sub-categories (Ordnance Survey 2016). We therefore made slight changes to the official classification scheme. By separating the original "health and education" category into two independent categories and moving all "government and organisations" POIs out of "public infrastructure" to become a major category by themselves, a new 11-category classification scheme that fits our research purpose was generated (Table 10.1).

Table 10.1 The reclassified POI categories based on the Ordnance Survey POI classification scheme

Customised classification scheme	
01 Accommodation, eating and drinking 01 Accommodation 02 Eating and drinking	**06 Public infrastructures** 34 Infrastructure and facilities
02 Commercial services 03 Construction services 04 Consultancies 07 Contract services 05 Employment and career agencies 06 Engineering services 60 Hire services 08 IT, advertising, marketing and media services 09 Legal and financial 10 Personal, consumer and other services 11 Property and development services 12 Recycling services 13 Repair and servicing 14 Research and design 15 Transport, storage and delivery	**07 Manufacturing and production** 37 Consumer products 38 Extractive industries 39 Farming 40 Foodstuffs 41 Industrial features 42 Industrial products **08 Retail** 46 Clothing and accessories 47 Food, drink and multi-item retail 48 Household, office, leisure and garden 49 Motoring
03 Attractions 58 Bodies of water 16 Botanical and zoological 17 Historical and cultural 19 Landscape features 18 Recreational 20 Tourism	**09 Transport** 53 Air 59 Bus transport 57 Public transport, stations and infrastructure 54 Road and rail 55 Walking 56 Water
04 Sport and entertainment 22 Gambling 23 Outdoor pursuits 21 Sport and entertainment support services 24 Sports complex 25 Venues, stage and screen	**10 Education** 27 Education support services 31 Primary, secondary and tertiary education 32 Recreational and vocational education **11 Government and organisations** 33 Central and local Government
05 Education and health 26 Animal welfare 28 Health practitioners and establishments 29 Health support services	*Annotation*: **XX Major category code** xx Sub-category code

10.3.4 Significance of the POIs

Just as the First Law of Geography formulated by Tobler (1970) proclaims that "everything is related to everything else, but near things are more related than distant things", POIs and buildings of identical functions are likely to aggregate in the same neighbourhood. Based on this phenomenon, Krüger et al. (2013) and Polisciuc (2015) used the quantity of POIs in a place or bounding area to explain the major semantic meanings of places in a simple manner. The significance of one type of POI (of major category I) to the area ST-ROI (j) they fall in can be expressed as the following (Eq. 10.1):

$$SC_{I,j} = f_{I,j} = \frac{count_{I,j}}{\sum_k count_{k,j}}$$

(10.1)

where $count_{I,j}$ is the number of major category I POIs in ST-ROI j, and $\sum_k count_{k,j}$ is the total number of all POIs in ST-ROI j. $f_{I,j}$ is therefore the raw frequency of the major category I POIs among all POIs in ST-ROI j.

Quantities of different categories of POIs vary dramatically in urban space, however. For instance, a large number of iconic public telephones and red pillar mailboxes can be found throughout London, but they have a relatively small influence on the meaning or function of an area. On the contrary, if there is only one museum in the entire city, the influence of this museum upon the local region where it is located should be magnified to outrank the many telephone boxes nearby. Hence, directly using the quantity of POIs for semantic enrichment is not enough. The bias caused by unbalanced quantities of different POIs should be subdued. An example in our study shows that 33% of the POIs in ST-ROI No. 10 belong to

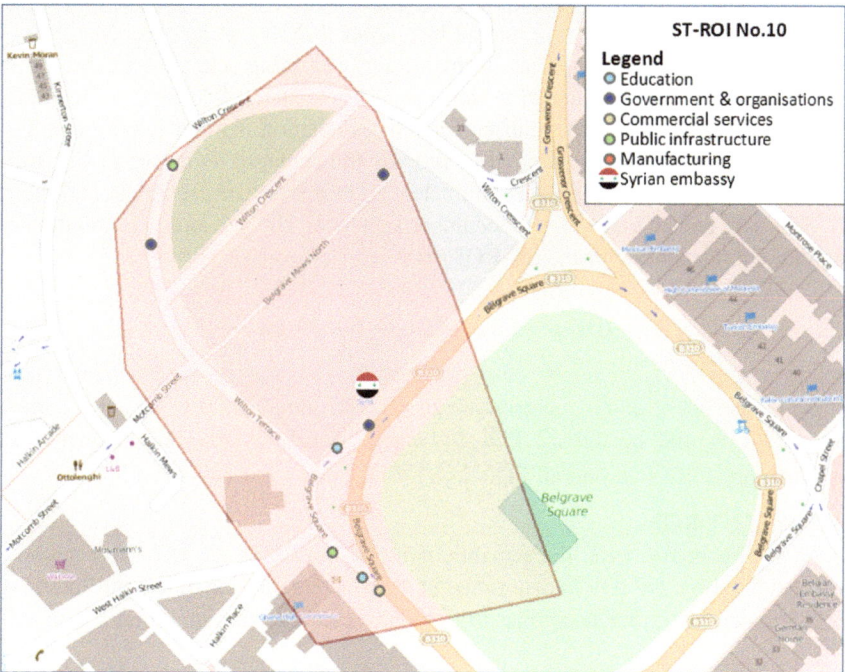

Fig. 10.2 Summarising the Points of Interest (POIs) located within the expanded convex hull of ST-ROI No. 10. The POIs' categories are represented by different colours and the difference the weighting process made for the 11 categories is shown on the right

'public infrastructure' category (see Fig. 10.2). As we know, public infrastructures, such as electricity poles and traffic lights, can be found in large numbers throughout the city and should be considered as less significant POIs in the semantic enrichment process.

A similar case can be found in text mining studies where article words like "the" and "a" appear far more frequently than the truly meaningful words in most sentences. The significance (semantic contribution) of a given word in one sentence increases proportionally to the number of times this word appears in the sentence, but is offset by the frequency of the word in the whole context. Likewise, the more sibling POIs (i.e. POIs that belong to the same major category) fall into an ST-ROI's expanded convex hull, the more impact they have on a place; however, the more sibling POIs exist in the entire district or other places, the less impact the POIs should have on the current place. In other words, POIs distributed ubiquitously are less relevant to the meaning of the local area.

In information retrieval and text mining studies, Term Frequency–Inverse Document Frequency (TF–IDF) is designed to measure the semantic contribution (weighting factor) of a word to the meaning of the sentence it falls in (Salton and Buckley 1988). TF–IDF can downplay the semantic contribution of a word if it appears everywhere in the entire article. Inspired by its function in semantic analysis of articles and documents, we introduce TF–IDF method to reweight and readjust the significance (semantic contribution) of different categories of POIs to each ST-ROI so that the negative effect of dominant insignificant POIs can be eliminated.

In our case of study, a double normalisation weighting scheme was chosen for term frequency (TF) calculation and a classic inverse document frequency (IDF) weighting scheme was chosen for IDF calculation, so that the original TF–IDF for text mining was amended to process the semantic POIs, as in Eq. 10.2. Every sub-category of POIs was equivalent to one word in the article and every ST-ROI was considered to be a document sentence. The semantic contribution of the POI sub-category i to the ST-ROI j is:

$$\text{TF–IDF}_{i,j} = \left(0.5 + 0.5\frac{f_{i,j}}{\max_i f_{i,j}}\right) \cdot \log\frac{N}{count_i} \tag{10.2}$$

where j is the set of points representing all POIs in one ST-ROI's expanded convex hull and i is one of the 52 POI sub-categories. $f_{i,j}$ is the raw frequency of sub-category i POIs among all POIs in ST-ROI j and $f_{i,j} = \frac{count_{i,j}}{\sum_k count_{k,j}}$. N is the total number of ST-ROIs.

As shown in Eq. 10.3, the semantic contribution of the 52 POI sub-categories was weighted by TF–IDF and summed up to generate the semantic contribution of the 11 major categories according to the major category to which they belonged.

$$\text{SC}_{I,j} = \sum_{i \in I} \text{TFIDF}_{i,j} \Big/ \sum_{I=1}^{11} \sum_{i \in I} \text{TFIDF}_{i,j} \tag{10.3}$$

where I is one of the 11 major categories. One major POI's semantic contribution to an ST-ROI is the normalised sum of the semantic contribution of all the sub-categories within a major category.

The expanded convex hull area of ST-ROI No. 10, an embassy outside Camden (Fig. 10.1), is presented in Fig. 10.2. This ST-ROI was generated by many police officers standing in front of the Syrian embassy keeping protestors out of the building. The simplest way of inferring the semantic meaning of a place is to directly use the quantity of sibling POIs as their semantic contribution index in an area. Figure 10.2 shows that the less significant "public infrastructure" POIs make up for a considerable portion of the total number of POIs in ST-ROI No. 10. Therefore, the semantic contribution of public infrastructures in ST-ROI No. 10 calculated by count is relatively high. Obviously, however, the police officers were not interested in the "public infrastructures" in that area. Quantity of public facilities did not translate into significant semantic contribution in this case. Therefore, Table 10.2 shows that the impact of the ubiquitous "public infrastructures" had been significantly weakened after the TF–IDF weighting process, from 0.333 to 0.043. In contract, small-size categories, such as educational POIs, were emphasised and the weight of governmental POIs were reinforced after the significance is recalculated with TF–IDF, corresponding to the common sense that this is an embassy area.

Table 10.3 shows the TF–IDF weights of the 11 categories of POIs in the 28 ST-ROIs. The weights are added to officers' staying time allocations (space time profiles) on ST-ROIs generate offices' semantic profiles (SP) in the next step.

Table 10.2 The semantic contribution weights of different categories of POIs in ST-ROI No. 10 calculated with Eqs. 10.1 and 10.3

	SC by count	TF-IDF weighted SC
Accommondation, eating, drinking	0	0
Commercial services	0	0
Attractions	0	0
Sport & entertainment	0	0
Health	0	0
Public infrastructure	0.3333	0.0434
Manufacture & production	0	0
Retail	0	0
Transport	0	0
Education	0.1667	0.3299
Government & organisations	0.5	0.6267

Table 10.3 The TF–IDF weighted semantic contribution of different categories of POIs in each ST-ROI in Camden

ST-ROI No.	1	2	3	4	5	6	7	8	9	10	11	12	13	14
Accommodation, eating and drinking	0.0230	0.1004	0.0964	0.1248	0.0271	0.0192	0	0.1040	0.0870	0	0	0.0257	0.1204	0.0856
Commercial services	0.1657	0.2751	0.2494	0.2533	0.3181	0.2806	0.2570	0.2712	0.2482	0	0.3890	0.7276	0.1957	0.2698
Attractions	0.2664	0.0103	0.0128	0.0238	0.0501	0.0711	0.1412	0.0136	0	0	0.4033	0.1488	0.0083	0.0142
Sport and entertainment	0	0.1022	0.0962	0.0640	0.0294	0.0416	0	0.0828	0.0993	0	0	0	0.0756	0.0746
Education and health	0.0907	0.0628	0.0666	0.0266	0	0	0	0.0478	0.0438	0	0	0.0338	0.0293	0.0533
Public infrastructure	0.0310	0.0233	0.0269	0.0134	0.0137	0.0162	0.0103	0.0203	0.0201	0.0434	0.0470	0.0087	0.0165	0.0206
Manufacturing and production	0.1179	0.0415	0.0234	0.0573	0.0456	0.0647	0	0.0441	0	0	0	0	0	0.0413
Retail	0.1677	0.2855	0.3304	0.3514	0.1476	0.1950	0	0.3135	0.4164	0	0	0.0257	0.4586	0.3446
Transport	0.0796	0.0518	0.0576	0.0514	0.0228	0	0	0.0558	0.0578	0	0.1607	0.0297	0.0423	0.0534
Education	0.0580	0.0317	0.0277	0.0248	0.0517	0.0733	0.1561	0.0298	0.0274	0.3299	0	0	0.0367	0.0357
Government and organisations	0	0.0153	0.0126	0.0092	0.2940	0.2384	0.4355	0.0172	0	0.6267	0	0	0.0166	0.0070

ST-ROI No.	15	16	17	18	19	20	21	22	23	24	25	26	27	28
Accommodation, eating and drinking	0.2339	0	0.1895	0	0.2471	0.1032	0.0573	0.1626	0.2867	0	0.1448	0	0	0.0439
Commercial services	0.3543	0.7264	0.3659	0	0.2438	0.2509	0	0.5179	0.1193	0.8689	0.4040	0.5718	0	0.2244
Attractions	0.0187	0	0.0205	0.1525	0	0.0159	0	0	0.0702	0	0	0	0	0.0290
Sport and entertainment	0.0438	0	0	0.0751	0.0833	0.1043	0.2611	0	0.0779	0	0.0872	0	0	0.0642
Education and health	0.0133	0	0.0145	0	0.0324	0.0525	0.1504	0.0184	0.0497	0	0.0280	0	0	0.0082
Public infrastructure	0.0306	0	0.0112	0.0111	0.0166	0.0207	0.0193	0.0047	0.0765	0	0.0287	0.0411	0	0.0084
Manufacturing and production	0.0792	0	0.0866	0	0	0.0332	0	0.0932	0	0	0.1079	0	0	0.0832
Retail	0.2145	0	0.2125	0	0.0861	0.3385	0	0.0921	0.3197	0	0.1451	0.1811	1	0.4177
Transport	0.0116	0.2736	0.0741	0	0.2343	0.0346	0.1320	0.1109	0	0.1311	0.0542	0.2060	0	0.1051
Education	0	0	0	0.1685	0	0.0349	0.3801	0	0	0	0	0	0	0.0158
Government and organisations	0	0	0.0252	0.5929	0.0563	0.0112	0	0	0	0	0	0	0	0

10.3.5 Semantic Profiling and Hierarchical Clustering

In our previous work, users' behaviour profiles were generalised based on their time budget allocation (i.e. space time profiles) in the ST-ROIs. In this way, the users' inclinations in different time periods and spatial locations were discovered. Figure 10.3 is an example of three officers (A, B and C) allocating their individual working hours differently in the 28 ST-ROIs; however, this method did not consider semantic meaning regarding the generated ST-ROIs. Spending time in two different places does not necessarily indicate any differences in visit purposes because the two places may have similar functions or semantic meanings even though they are far apart. Moreover, the previous method cannot work in large cities because more and more ST-ROIs will be detected as the study area expands dramatically, and the difference between users' time allocation profiles will be erased. By summarising ST-ROIs into a limited number of categories according to their semantic and functional meanings, the time budget allocation analysis can be translated into semantic profiles that demonstrate the time spent on different semantic places instead of meaningless locations.

Most of the officers on duty are foot patrol officers (FP), community support officers (CSO) and senior officers (SO). The time that different types of officers allocate to ST-ROIs can be very different, because the different tasks they are required to undertake are determined by their types. The time allocation profile of three typical police officers can be seen in Fig. 10.3; the identity call signs of the officers have been encrypted for security's sake.

In order to understand how officers performing different roles on patrol allocate their attention across different functional areas and activities, the TF–IDF weights of the ST-ROIs were added to the analysis. One officer's staying time in a ST-ROI

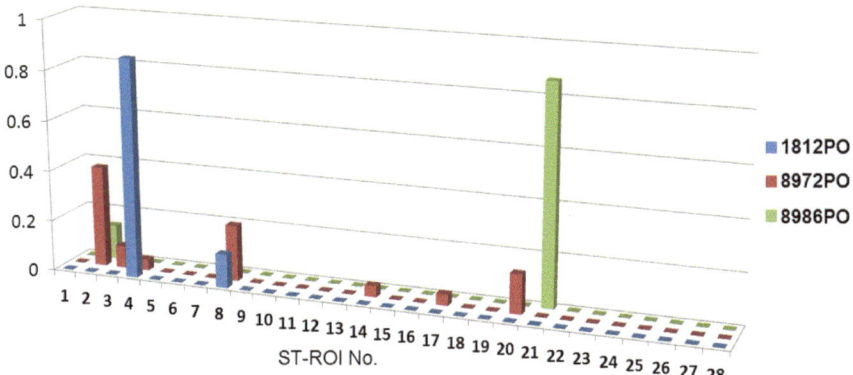

Fig. 10.3 Three police officers' time allocation in 28 ST-ROIs

was assigned to the 11 categories of POI according to each POI's semantic contribution to that specific ST-ROI, as Eq. 10.4 describes. This process turns space time profiles (i.e. time allocations across ST-ROIs) into semantic profiles (i.e. allocation across functional areas), so that the differences in higher activity levels can be revealed.

$$SP_{o,I} = P_{o,j} \cdot SC_{I,j} \tag{10.4}$$

where $P_{o,j}$ is the space time profile of officer o, and $SP_{o,I}$ is the semantically-weighted profile of officer o in the 11 POI major categories.

For instance, officer 8986PO spend 90% of his/her staying time in ST-ROI No. 21 and 10% staying time in ST-ROI No. 2. According to the semantic weights in Table 10.3, 2.75% $(0.2751 * 10\% + 0 * 90\%)$ of 8986PO's total active time is assigned to places of commercial services, whereas 34.5% $(0.0317 * 10\% + 0.3801 * 90\%)$ of 8986PO's total active time is assigned to educational venues. Figure 10.4 displays the comparison of the three chosen officers after their space time profiles are turned into semantic profiles by Eq. 10.4.

It is worth noting that the time allocations to ST-ROIs of officers "1812PO", "8972PO" and "8986PO" were by no means the same (Fig. 10.3). After interpreting the behaviours of the officers with POI impacts, however, the profiles of officer "1812PO" and officer "8972PO" became quite similar to each other semantically (Fig. 10.4). This is because they have visited semantically similar places, despite the locations they have visited are different. Their common interests in retail and commercially-related areas were revealed, whereas semantically interpreting the profile of officer "8986PO" made it prominently distinct from the other two officers. This shows the new method's capability to find users with similar behaviours, despite the fact that the places in which they are active may be spatially far apart.

After profiling the users (officers), the pairwise differences among the officers' TF–IDF weighted profiles could be quantified with Jessen Shannon Distance (JSD) (Lin 1991) and similar profiles could be grouped together as a result of the

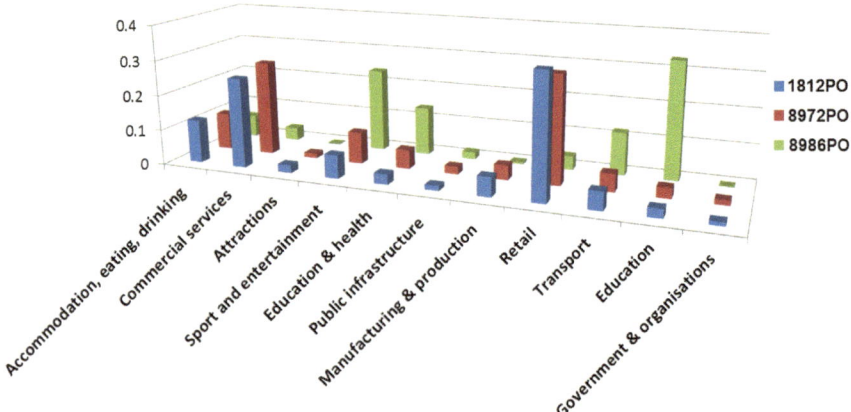

Fig. 10.4 The summarised user semantic profiles of three police officers, i.e. the officers' time allocation to different semantic places

hierarchical clustering method. Equation 10.5 shows how the JSD value between semantic weight profiles of two people, SP.p and SP.q, is calculated.

$$
\begin{aligned}
\mathrm{JSD}\,(\mathrm{SP.p}|\mathrm{SP.q}) = {} & \frac{1}{2} \sum_i \mathrm{SP.p}(t) \ln \frac{2 \cdot \mathrm{SP.p}(t)}{\mathrm{SP.p}(t) + \mathrm{SP.q}(t)} \\
& + \frac{1}{2} \sum_i \mathrm{SP.q}(t) \ln \frac{2 \cdot \mathrm{SP.q}(t)}{\mathrm{SP.p}(t) + \mathrm{SP.q}(t)}
\end{aligned}
\tag{10.5}
$$

Figures 10.5 and 10.6 show the hierarchical clustering results without and with the consideration of semantic meaning and the significance of the places, respectively. With the space time profiles in previous method (Shen and Cheng 2016), the officers are divided into 8 groups, as suggested by the Dunn index (Dunn 1973). With the newly proposed semantic profiles, the officers are divided into 5 groups. The call signs of officers have been encrypted and the three exampled officers are marked with dashed rectangles. In Fig. 10.5, the three example officers are grouped into three different groups because of their differences in space time profiles. In contrast, Fig. 10.6 shows that the new method can find the semantic similarities between officers "1812PO" and "8972PO", despite their location differences. The new method also generated clearer and simpler grouping results than the old one. This is because there are a smaller number of summarised semantic categories than the actual number of the extracted ST-ROIs.

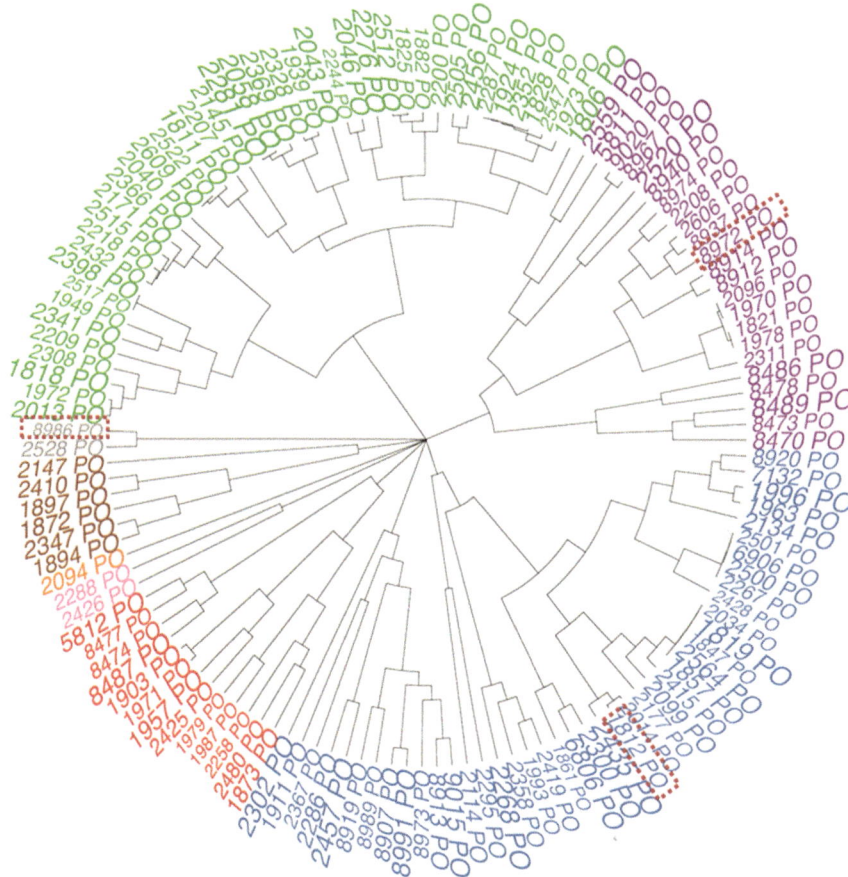

Fig. 10.5 Taxonomy tree showing the clustering results of previous work purely based on officers' space time profiles (time allocation to ST-ROIs) (Shen and Cheng 2016)

10.4 Extended Case Study

Greater London consists of 32 boroughs (local authority districts), each of which is assigned a Borough Operational Command Unit (BOCU) of Metropolitan Police. All BOCUs provide police officers (regulars and specials) who are responsible for patrolling and responding to emergencies.

We have showcased how the new method works within one borough and how different it is from the previous method above. Here, to prove that the new method improves its capability to process larger scale data, we implemented the algorithms in parallel with police foot patrol activities in all 32 BOCUs, based on a new version of the APLS dataset. Thanks to the hardware upgrades of the Metropolitan

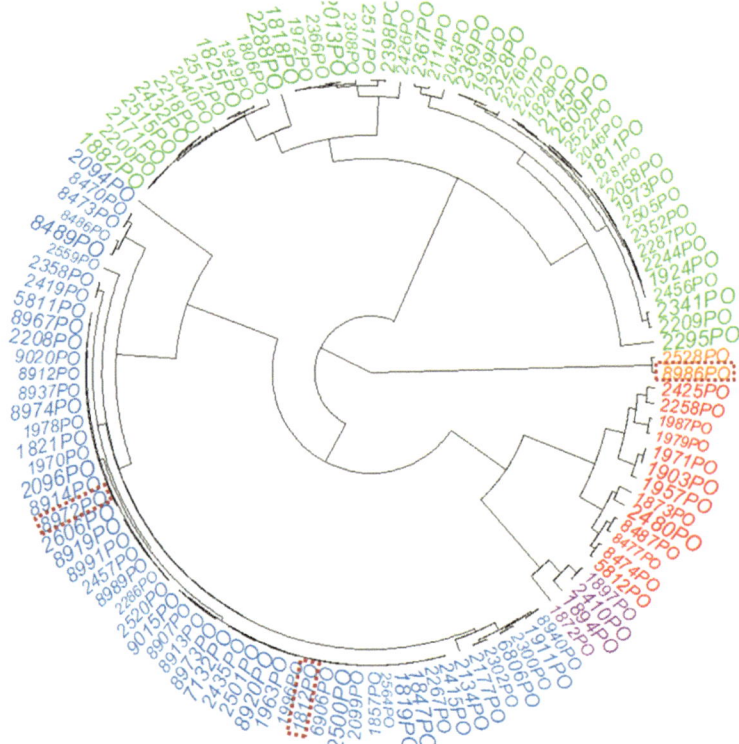

Fig. 10.6 Taxonomy tree showing the clustering results based on semantic profiles

Police, the location accuracy has improved and sampling rate of the GPS devices is now updated every 5 min.

ST-DBSCAN was first used on all metropolitan police patrol activities in all BOCU areas in one month in 2015. For simplicity of visualisation, we demonstrated the results of the three inner boroughs (City of Westminster, Islington and Camden) in central London. Figure 10.7 shows all 54 ST-ROIs generated in the extended case study area. For security reasons, we cannot label those ST-ROIs, though we choose 5 places that people are familiar with in the next step.

The second step in the extended case study was to import the POI data of the entire Greater London and enrich the semantic meaning of the 54 ST-ROIs. Table 10.4 shows the semantic weights of POIs in the five ST-ROIs shown in Fig. 10.7 as examples. It shows that the TF–IDF results are in line with the citizens' common impression on the meaning of the places. These weights were used to explain the semantic meanings of the detected ST-ROIs and turn the officers' time allocation to ST-ROIs into time allocation to semantic places, in order to generate the semantic profiles according to Eq. 10.3.

Table 10.4 The TF–IDF semantic weights and names of five chosen ST-ROIs in the extended case study

ST-ROI ID	2	6	7	27	34
Name of the place	Backingham Palace	Soho	Trafalgar Square	Camden Station	White Hall
Accommodation, eating, drinking	0.0000	0.3233	0.0642	0.1041	0.1375
Commercial services	0.0487	0.1676	0.1910	0.2370	0.1818
Attractions	0.7438	0.0375	0.2413	0.0119	0.0156
Sport & entertainment	0.0000	0.1339	0.0000	0.1047	0.0288
Health	0.0000	0.0411	0.0000	0.0812	0.0000
Public infrastructure	0.0233	0.0387	0.0644	0.0236	0.0268
Manufacture & production	0.0000	0.0426	0.0203	0.0267	0.0000
Retail	0.0523	0.1412	0.0419	0.3027	0.0000
Transport	0.0000	0.0332	0.2551	0.0375	0.0357
Education	0.0000	0.0296	0.0300	0.0534	0.0000
Government & organisations	0.1319	0.0114	0.0918	0.0170	0.5738

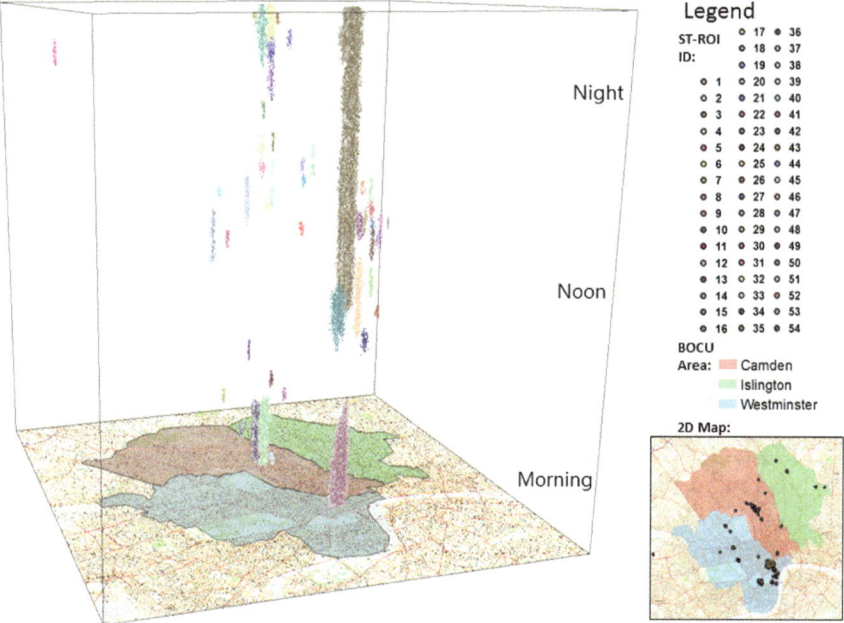

Fig. 10.7 The 54 ST-ROIs in three BOCU areas chosen for demonstration

After the semantic profiles were grouped via the JSD-based hierarchical clustering method, officers with similar activity patterns across all boroughs could be detected, even if they never belonged to the same BOCU area branch. 54 ST-ROIs are detected in the three boroughs and even more will be generated if method is to be applied to the entire London. The drastically increasing number of ST-ROIs means that the time allocation profile will contain much more variables in the previous methodological framework (Shen and Cheng 2016), which will lead to the 'curse of dimensionality' (Bellman 1961) in the following clustering process. By applying the newly proposed method in this work, the number of dimensions of the semantic profile is limited to 11 (the number of POI major categories) and the problem of 'curse of dimensionality' in large study areas can be avoided.

As the Dunn index test suggested, the optimal group number should be five in the hierarchical clustering for the extended case study. There are more than 600 active officers patrolling in these three boroughs, which made the visualisation of the hierarchical results shown in Figs. 10.5 and 10.6 impossible to present properly on the printed page. The average semantic profile of each officer group was summarised to show the representative pattern of their activities (Fig. 10.8). It shows that the focus of officers on different semantic places varied greatly. Officers in Group 1 allocated their time more evenly than did others and paid more attention to commercial and retail streets. Group 2 preferred to stay near tourist attractions, whereas Group 3 focused on sport and entertainment events and Group 5 patrolled around both governmental and public infrastructures. Group 4 spend most of their time nearby hospitals and has much less activities than other groups. This demonstrates that the activity patterns of police officers show clear differences when the semantic meaning of places is brought into the profile clustering process, and each group has its own major interest.

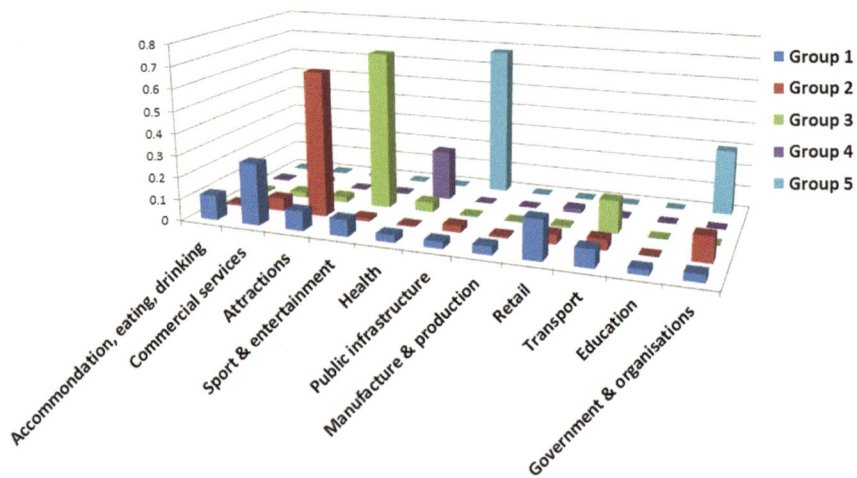

Fig. 10.8 ST-ROIs in three BOCU areas chosen for demonstration

10.5   Conclusion and Discussions

New datasets of time-series locations have enabled the study of behaviours that conventional studies cannot process, due to the lack of advanced status logging approaches. In this research, we improved our previous work (Shen and Cheng 2016) by introducing new concept that '*what place, when and how long you stay is who you are*' to focus on place than locations. Methodologically, the semantics and significance of the places are extracted and measured based upon POIs in cities, so that the new method is able to extend the behaviour grouping to a larger spatial scale and generate more meaningful results. We used police foot patrol data as a case study to represent kindred location-based applications.

The framework further extends the traditional ideas of time budget allocation in behavioural studies and existing spatial-location-based user similarity definitions to a semantic explanation of people visiting places. It can profile the activity patterns of people according to space, time and semantic aspects by defining the JSD behavioural similarity metric, which is closer to people's place visiting purposes in reality. Furthermore, after determining what the place is about semantically, the pattern differences of individuals' activities are better explained. The developed method can be applied to other massive GPS movement datasets, such as volunteer-collected cyclist and taxi data, and can detect similar communities and anomalies within them. It can also work with location-based check-in services to support friend recommendation applications in location-based social networks.

Further work will turn this methodological framework into a street network-based version in order to further improve the accuracy of the estimated police patrol routes and visited locations, and to further adapt the method to the urban street environment. To this end, map-matching algorithms and a network-based space-time clustering method will be developed.

Acknowledgements This work is part of the project—Crime, Policing and Citizenship (CPC): Space-Time Interactions of Dynamic Networks (www.ucl.ac.uk/cpc), supported by the UK Engineering and Physical Sciences Research Council (EP/J004197/1). The data provided by Metropolitan Police Service (London) is highly appreciated. The second author's Ph.D. research is funded by the China Scholarship Council (CSC). The CSC is a non-profit institution with legal person status affiliated with the Ministry of Education in China.

References

Alvares, O., et al. (2007). A model for enriching trajectories with semantic geographical information. In *Proceedings of the 15th Annual ACM International Symposium on Advances in Geographic Information Systems*. Article No. 22. New York: ACM. https://doi.org/10.1145/1341012.1341041.

Alves, A. O., Rodrigues, F., & Pereira, F. C. (2011). Tagging space from information extraction and popularity of points of interest. In *International Joint Conference on Ambient Intelligence* (pp. 115–125). Berlin, Heidelberg: Springer.

Andrew, A. (1979). Another efficient algorithm for convex hulls in two dimensions. *Information Processing Letters, 9*(5), 216–219.

Andrienko, G., Andrienko, N., Bak, P., Keim, D., & Wrobel, S. (2013). *Visual analytics of movement.* Berlin: Springer Science & Business Media.

Bellman, R. E. (1961). *Adaptive control processes: A guided tour.* Princeton: Princeton University Press.

Birant, D., & Kut, A. (2007). ST-DBSCAN: An algorithm for clustering spatial-temporal data. *Data & Knowledge Engineering, 60,* 208–221. https://doi.org/10.1016/j.datak.2006.01.013.

Braun, et al. (2010). Collaborative semantic points of interests. *Extended semantic web conference* (pp. 365–369). Berlin, Heidelberg: Springer.

Damiani, M. L., Silvestri, C., & Bertino, E. (2011). Fine-grained cloaking of sensitive positions in location-sharing applications. *IEEE Pervasive Computing, 10*(4), 64–72.

Dunn, J. C. (1973). A fuzzy relative of the ISODATA process and its use in detecting compact well-separated clusters. *Journal of Cybernetics, 3*(3), 32–57.

Giannotti, F., Nanni, M., Pinelli, F., & Pedreschi, D. (2007). Trajectory pattern mining. In *Proceedings of the 13th ACM SIGKDD International Conference on Knowledge Discovery and Data Mining* (pp. 330–339).

Goodchild, M. (2015). Space, place and health. *Annals of GIS, 21*(2), 97–100.

Güting, R. H., de Almeida, T., & Ding, Z. (2006). Modelling and querying moving objects in networks. *International Journal on Very Large Data Bases, 15*(2), 165–190.

Karli, S., & Saygin, Y. (2009). Mining periodic patterns in spatio-temporal sequences at different time granularities. *Intelligent Data Analysis, 13*(2), 301–335.

Krüger, R., Thom, D., & Ertl, T. (2015). Semantic enrichment of movement behaviour with foursquare—A visual analytics approach. *IEEE Transactions on Visualization and Computer Graphics, 21*(8), 903–915.

Krüger, R., Thom, D., Wörner, M., Bosch, H., & Ertl, T. (2013). Trajectory lenses—A set-based filtering and exploration technique for long-term trajectory data. *Computer Graphics Forum, 32,* 451–460.

Lee, J. G., Han, J., & Whang, K. Y. (2007). Trajectory clustering: A partition-and-group framework. In *Proceedings of the 2007 ACM SIGMOD International Conference on Management of Data* (pp. 593–604).

Li, Z., Ding, B., Han, J., Kays, R., & Nye, P. (2010). Mining periodic behaviours for moving objects. In *Proceedings of the 16th ACM SIGKDD International Conference on Knowledge Discovery and Data Mining* (pp. 1099–1108).

Liao, L., Patterson, D. J., Fox, D., & Kautz, H. (2006). Building personal maps from GPS data. *Annals of the New York Academy of Sciences, 1093*(1), 249–265.

Lin, J. (1991). Divergence measures based on the Shannon entropy. *IEEE Transactions on Information Theory, 37*(1), 145–151. https://doi.org/10.1109/18.61115.

Miller, R., & Stout, F. (1988). Efficient parallel convex hull algorithms. *IEEE Transactions on Computers, 37*(12), 1605–1618.

Ordnance Survey. (2016). Points of interest [online]. Available from: https://www.ordnancesurvey.co.uk/business-and-government/products/points-of-interest.html.

Palma, A. T., Bogorny, V., Kuijpers, B., & Alvares, L. O. (2008). A clustering-based approach for discovering interesting places in trajectories. In *Proceedings of the 2008 ACM symposium on Applied computing* (pp. 863–868).

Parent, C., Spaccapietra, S., Renso, C., Andrienko, G., Andrienko, N., Bogorny, V., Damiani, M. L., Gkoulalas-Divanis, A., Macedo, J., Pelekis, N., & Theodoridis, Y. (2013). Semantic trajectories modeling and analysis. *ACM Computing Surveys (CSUR), 45*(4), 42.

Polisciuc, E. (2015). Understanding urban land use through the visualization of points of interest. In *Proceedings of the Fourth Workshop on Vision and Language* (pp. 51–59).

Reumers, S., Liu, F., Janssens, D., Cools, M., & Wets, G. (2013). Semantic annotation of global positioning system traces: Activity type inference. *Transportation Research Record: Journal of the Transportation Research Board, 2383,* 35–43.

Salton, G., & Buckley, C. (1988). Term-weighting approaches in automatic text retrieval. *Information Processing and Management, 24*(5), 513–523.

Shaw, S. L., Tsou, M. H., & Ye, X. (2016). Human dynamics in the mobile and big data era. *International Journal of Geographical Information Science, 30*(9), 1687–1693.

Shen, J., & Cheng, T. (2016). A framework for identifying activity groups from individual space-time profiles. *International Journal of Geographical Information Science, 30*(9), 1785–1805. https://doi.org/10.1080/13658816.2016.1139119.

Slocum, T. A., McMaster, R. M., Kessler, F. C., Howard, H. H., & Mc Master, R. B. (2008). Thematic cartography and geographic visualization.

Spaccapietra, S., Parent, C., Damiani, M. L., de Macedo, J. A., Porto, F., & Vangenot, C. (2008). A conceptual view on trajectories. *Data & Knowledge Engineering, 65*(1), 126–146.

Tobler, W. (1970). A computer movie simulating urban growth in the Detroit region. *Economic Geography, 46*(2), 234–240.

Tsou, M. H. (2015). Research challenges and opportunities in mapping social media and big data. *Cartography and Geographic Information Science, 42*(1), 70–74. https://doi.org/10.1080/15230406.2015.1059251.

Yan, Z., Chakraborty, D., Parent, C., Spaccapietra, S., & Aberer, K. (2013). Semantic trajectories: Mobility data computation and annotation. *ACM Transactions on Intelligent Systems and Technology (TIST), 4*(3), 49.

Ying, J. J. C., Lee, W. C., & Tseng, V. S. (2013). Mining geographic-temporal-semantic patterns in trajectories for location prediction. *ACM Transactions on Intelligent Systems and Technology (TIST), 5*(1), 2.

Zhao, X., & Xu, W. (2011). Clustering spatio-temporal trajectories based on compression of interesting places. *Journal of Beijing Jiaotong University, 3*, 10.

Zheng, Y., Chen, Y., Xie, X., & Ma, W. Y. (2009). GeoLife2. 0: a location-based social networking service. In *Mobile Data Management: Systems, Services and Middleware* (pp. 357–358).

Zimmermann, M., Kirste, T., & Spiliopoulou, M. (2009). Finding stops in error-prone trajectories of moving objects with time-based clustering. *Intelligent Interactive Assistance and Mobile Multimedia Computing* (pp. 275–286). Berlin, Heidelberg: Springer.

Author Biographies

Tao Cheng is a Professor in GeoInformatics, the Founder and Director of SpaceTimeLab for Big Data Analytics (www.ucl.ac.uk/spacetimelab) at University College London. She is a world-leading expert in big data analytics for smart cities. Her research interests span network complexity, geocomputation, spatio-temporal analytics with applications in transport and mobility, security and safety, public health, business intelligence and hazard prevention. She has over 200 publications and secured over £12M research grants, working with government and industrial partners including Transport for London (www.ucl.ac.uk/standard), the London Metropolitan Police Service (www.ucl.ac.uk/cpc/) and Public Health England (http://www.cdrc.ac.uk), Arup and Bosch, to name a few.

Jianan Shen is a Ph.D. candidate in SpaceTimeLab for Big Data Analytics (www.ucl.ac.uk/spacetimelab), UCL. He acquired his BEng in Information Engineering in China's National University of Defense Technology with first-class honours (2013) and thereafter undertook his Ph.D. programme in the UK with full scholarship support from the China Scholarship Council. During his study in the UCL, he joined in the Crime, Policing and Citizenship Project (www.ucl.ac.uk/cpc/) sponsored by the EPSRC from 2013 to 2016, working on the space time pattern analysis of police urban patrol behaviours. His research interests include Spatio-temporal Movement Analysis, Human Dynamics and Semantic Enrichment of Places.

Chapter 11
Open Source Social Network Simulator Focusing on Spatial Meme Diffusion

Xinyue Ye, Lanxue Dang, Jay Lee, Ming-Hsiang Tsou and Zhuo Chen

11.1 Introduction

In today's geospatially connected world, data-driven computing and analytics have become increasingly essential to understand coupled human–environment systems (Wang 2016). Indeed, the complexities of such systems and their connectivity at various spatial and temporal scales have posed daunting challenges to effective solutions to a variety of sustainable development issues. Due to the rapid progress of spatial data science and open social data, opportunities have been presented for human-environment interaction research from a very finer scale to global scale.

Social media has been gaining a spectacular popularity in the past several years. With the proliferation of social media platforms such as Twitter, Facebook, and Instagram, information generated and disseminated from these outlets has become an important part of our everyday lives. A meme is an idea, behavior, or style that spreads from person to person (Chesterman 2016). The dynamics of meme has been facilitated by the wide adoption of social media apps, along with the widespread of mobile devices. Social media messages contain a wealth of location information, which can help us understand some of the phenomena associated with the geographic location and reveal the law hidden behind the phenomenon. Location-based social media data offer a golden opportunity to examine spatial social dynamics at multiple scales. Particularly noteworthy is that users can post the latest stories, ideas, opinions and news about what users consider are exciting or important in

X. Ye (✉) · J. Lee · Z. Chen
Department of Geography, Kent State University, Kent, USA
e-mail: xye5@kent.edu

L. Dang
Department of Computer Science, Henan University, Kaifeng, China

M.-H. Tsou
Department of Geography, San Diego State University, San Diego, USA

© Springer International Publishing AG, part of Springer Nature 2018
S.-L. Shaw and D. Sui (eds.), *Human Dynamics Research in Smart and Connected Communities*, Human Dynamics in Smart Cities,
https://doi.org/10.1007/978-3-319-73247-3_11

geographical and social environment. For instance, social media has become an important emergency information and communication backbone where individuals and organizations request and share information for disaster relief within and outside the affected area (Wang et al. 2016).

As Wang and Ye (2017) stated, "due to the capability of capturing human activities, social sensing techniques featured by various big data sources such as social media data and movement data are gaining increasing attention from geographic information scientists and domain scientists". It is important to discover, track, summarize, and even predict popular topics and events occurring in the social network in the space-time context (Shaw et al. 2016). At the same time, it is very useful that a series of "what if" scenarios can be developed to estimate the meme diffusion. However, spatial social scientists have been slow to adopt and implement new methods for social media data analysis due to the lack of open source software packages, which become a major impediment to the promotion of human dynamics research. The availability and widespread use of source codes will play a critical role in the adoption of new perspectives and ideas enhancing spatial social network analytics (Ye and Rey 2013). The proposed Open Source Social Network Simulator implements the methodological advances in an open source environment of Python for exploring spatial meme diffusion, using twitter data as the case study. Synthesis of spatial and social network analytics has contributed to our understanding of geography of social media such as the spatiotemporal effect in the information diffusion process. More toolkits are needed to interface the open source revolution and human/socioeconomic dynamics analysis seeking cross-fertilization between these two fast-growing communities (Ye 2017). The methods are built in open source environments and thus are easily extensible and customizable. The open source movement can also facilitate the explosion of the social media analytics routines by increasingly easier development processes with powerful scripting language environments.

The motivation initiating the development of this toolkit is discussed in the next section of literature review. Section 11.3 provides an overview of the design, implementation, and functionality of Social Network Simulator. A selection of its analytical capabilities are then demonstrated that highlight the strength and extensibility of the package in the follow-up section. The article is summarized with a roadmap of future plans.

11.2 Literature Review

An integrated open source statistical and visual toolkit is needed to support the analysis of multiple dimensions of social media data. Extensive studies with significant societal impacts have been recently reported based on location-based social media data. Hundreds of millions of social media users, sometimes referred to as "citizen sensors", share the opinions, experiences and observations of surrounding environments via social media, and at the same time generate a huge amount of social media data sets (e.g., text messages, photos, videos, and structure of social network). The inclusion of location or spatial dimensions in social media, or location-based social media (LBSM), blurs the interface between the cyberspace of social media and geographic space of the real world, and renders the social media as a promising lens to examine human behaviors and social dynamics. Our ways of examining social-spatial interactions are increasingly transformed by the development of more powerful computing technologies, emerging big and open data sources, and new perspectives on social-spatial processes (Shaw et al. 2016; Li et al. 2017).

There have been extensive studies on how information (media contents) diffuse and spread in a social communication network (Ye et al. 2018). In early studies, researchers have utilized and refined the disease-propagation SIR (Susceptible, Infected and Recovered) model in epidemiology (Bailey 1975) to simulate and understand the information spreading process (Newman et al. 2002). Many recent studies considered the spread of a piece of information through a social network as an innovation diffusion process, where two fundamental models, LT (Linear Threshold) model (Granovetter 1987) and IC (Independent Cascade) model (Goldenberg et al. 2001), are frequently used. A few studies introduced more generalized LT and IC models by relaxing their synchronization nature (Guille and Hacid 2012). Several other approaches, such as linear influence model (LIM) (Yang and Leskovec 2010), partial differential equations (PDE) (Wang et al. 2012), and regression based methods (Yang and Counts 2010), are proposed for predicting the spreading of information. Network structures (Weng et al. 2013), semantics/topics (Romero et al. 2013), and temporal-spatial properties (Liang et al. 2013), are also essential for information spreading.

The processes of meme diffusion over network communities can be modeled in several ways following the various approaches in the spatial diffusion studies since Hägerstrand (1967). Gregory and Urry (1985) provided a critique to Hägerstrand's models, noting that such models lack the ability to deal with spatial diffusion over intangible media such as social networks and the model's inability to deal with conflicts and resistance that may exist in the networks. Following this, Morrill et al. (1988) reviewed and discussed quantitative models of spatial diffusion processes by categorizing them as stochastic models or deterministic models. A stochastic model is one in which the elements include probability. This means that an observed spatial pattern of diffusion phenomena may be the result of forces that have a random component. Alternatively, a deterministic model does not allow for chance. With a deterministic model, the way a geographic phenomenon diffuses is according to certain fixed forms in deterministic models. Several spatial diffusion models were discussed in Morrill et al. (1988) that include mathematical variants of the Hägerstrand's model, epidemiology model, and spatial-temporal models. Besides adding spatial terms to classic statistics and using them to help detect and distinguish contagious and hierarchical diffusion processes, Allaway et al. (1994, 2003) added spatial and temporal lag as variables in event history analysis to describe contagious diffusion processes for how retail market areas evolved over time. Knoke (1982) applied event history analysis by incorporating the population size and spatial lag to illustrate the spread of local governing policies as hierarchical and expansion diffusion processes. Similar approaches can be seen in studies on diffusion of union formation and formation of political parties, same-sex marriage bans, spread of changes in income tax policies, and ethnic violence.

Researchers have been actively investigating the spatiotemporal effect in the information diffusion process across a wide range of disciplines. Baybeck and Huckfeldt (2002) investigated the spatial and temporal diffusion of political information within urban areas. Zhang et al. (2012) examined an information diffusion theory-based methodology for spatiotemporal risk assessment of natural disasters. Doo (2012) presented the activity-base social influence model based on activity enhanced heat diffusion kernel and a suite of activity influence rank based top k algorithms. Cao et al. (2012) proposed a visualization design, "Whisper", for tracing the process of information diffusion in social media in real time. The design highlights three major characteristics of diffusion processes in social media: the temporal trend, social-spatial extent, and community response of a topic of interest. Such social, spatiotemporal processes are conveyed based on a sunflower metaphor whose seeds are often dispersed far away.

Spread of information or ideas is not confined to either a localized geography or through a limited network of contacts (Lee et al. 2014). New opportunities emerge as we are able to model interactions between social and physical spaces at multiple spatial and temporal scales and with very fine granularity (Spitzberg 2014). Citizens produce, assemble, and diffuse a large variety of social media information for enhancing situational awareness especially during emergency management and disaster relief (Wang et al. 2016). Social media data are multi-dimensional. Besides analyzing spatial and temporal characteristics of social media data, some studies focused on mining the actual content of social media messages to improve knowledge about situations. Each tweet contains multiple fields such as user ID, post time, text, coordinates, and retweet relationship. Space, time, content, and network are all important attributes of social media data and should be fully used to gain insights into situational awareness (Wang et al. 2016). In Twitter, the hashtag may be considered as a surrogate of topics or events; however, in the general social, news and blog media, topics and events have to be extracted and recognized from the contents. Researchers have utilized the temporal and social dynamics to help discover and understand the rise and fall of emerging topics and events in social media (Wang et al. 2017). Identifying the influence spreaders in a network has become a major task for viral marketing, targeted advertisement, and even emergency response; these spreaders can help quickly and maximally spread a piece of information in a network environment. This problem is often referred to as an influence maximization problem in social network mining, with the goal of selecting a list of seed nodes, which can ensure the information/adoption reaching on average a maximal number of nodes (Kempe et al. 2003). In the meantime, there are studies aiming to identifying the top spreaders from the social media and network analysis (Romero et al. 2011). Finally, since misinformation and/or rumors can also be spread into the system, how to (quickly) identify and stop the spread the spread of misinformation is hence of considerable interest. Budak et al. (2011)

studied how to limit the spread of misinformation in social networks by using "limiting campaigns" to counteract the effect of misinformation. They proposed algorithms to minimize the efforts of a spread of misinformation.

11.3 Overview of Toolkit

Many efforts have been witnessed in the development of powerful packages for analyzing the structure and dynamics of social network/media data. For example, Gelphi is an open source interactive visualization and exploration platform investigating network and its properties. Pajek, an open source toolkit on the very large networks, can conduct more complicated analyses. Developer tools such as NetworkX and SNAP can be customized to calculate many computationally intensive metrics in the network data sets. However, these tools cannot simulate information propagation. Hence, our toolkit integrates network generator, network analysis, community detection, and information diffusion into one open source package to conduct computing and simulation on both artificial and real-world networks. Developed using the Python object-oriented scripting language, these modules can be combined to facilitate interactive programming and access to the many spatial analysis and network analysis libraries contained within Python.

This open source toolkit aims to explore the pattern of information diffusion over time on the social network, so it displays a complete process that starts with preparing a network and ends with demonstrating how information propagates over the network. There are four core modules in this toolkit, including Network Generator, Network Analysis, Community Detection, and Information Diffusion. The related functions are described in Table 11.1. A graphical user interface (GUI) integrates most of the analytical modules to support the interactive exploration of the spatial, temporal, and network dimensions of Meme Diffusion processes. Spatiotemporal visualization can be applied to intuitively reflect the complex process of information diffusion. These visualization techniques allow analysts to iteratively and interactively explore the dataset and thus gain deeper

Table 11.1 Modules and description of open source social network simulator

Modules	Function description
Network generator	Generate networks with different structures based on parameter settings
Network analysis	Examine the characteristics of networks or nodes
Community detection	Detect the distinct groups in a network
Information diffusion	Provides various algorithms to select seed nodes and opinion leaders to demonstrate information diffusion over network in the independent cascade model and the linear threshold model

understanding of the origination, propagation, and clustering of information. Scripts in this toolkit can also be customized for batch-oriented analyses and simulation.

11.3.1 Network Generator

Social network is where information propagates. It is noted that each type of network has different structure and related characteristics, such as average degree, network centrality, and average shortest path. There is a growing awareness that the network structure would affect the procedure of information dissemination (Bakshy et al. 2012). Centola (2010) examined how social networks might affect the spread of behavior, finding clustered-lattice networks are more effective for behavioral diffusion than random networks. This module can generate various network models, such as random network, full network, small-world network, and preference attachment network.

Random network can be generated by connecting any two randomly chosen nodes until a desired number of edges is achieved (Erdös and Rényi 1960). In the Full network, edges exist between any two nodes (Hanneman 2000). Both random network and full network are not the representative of the real world, since most networks are neither random nor regular. Small-world network is a sparse network featured with larger cluster coefficient, shorter average path length, and scalable entropy (Barrat and Weigt 2000). Watts and Strogatz (1998) developed a small-world network from a regular network by adjusting edges and properties of the original network. In a preferential attachment network, a few hubs are extremely well connected, while the rest are poorly linked (Barabási and Albert 1999). This module provides various network models to better understanding the differences among them and explore how these varieties might influence the procedure of information diffusion. More network models or modified models can be added to this module on demand because this tool is highly customizable in the Python environment.

11.3.2 Network Analysis

This module provides both a theoretical framework that conceptualizes the connection among individual actors and a group of analytic methods modeling these interactions. On one hand, this module delivers some metrics to measure social network, such as number of nodes, number of edges, average degree, average shortest path, modularity, and diameter (Scott 2012). On the other hand, the feature of a node can be described using the indicators such as betweenness centrality, closeness centrality, eigenvector centrality, degree, out degree, and in degree, identifying the relative importance of a node within a network (Newman 2008). All social networks are modeled using similar relational structure composed of nodes and edges, and are analyzed based on a set of mathematical methods grounded in graph theory (Newman 2003).

11.3.3   Community Detection

Community is an underlying structure representing how individual actors interact with each other in social network. Actors or nodes in the same community are more possibly influenced than those in different communities. This module contains three algorithms of community detection, including OSLOM (Lancichinetti et al. 2001), CNM (Clauset et al. 2004) and GN (Girvan and Newman 2002). OSLOM optimizes locally the statistical significance of communities. CNM is a hierarchical agglomeration algorithm for large-scale networks. GN algorithm detects the communities based on betweenness centrality. Users can explore which community structure is more efficient to propagate messages during the information diffusion through this module.

11.3.4   Information Diffusion

To simulate how information is disseminated over time through social network among nodes (users), this module implements independent cascade model (IC model) and linear threshold model (LT model). Independent cascade model is sender-centric, where a node (user) sends a message and all connected nodes has the same probability to receive the message. While linear threshold model is receiver-centric, where whether a node might be influenced depends on its neighbors with weights representing how influential it is. The focal node will be influenced if the sum of its active neighbors' weights reaches a threshold. This simulation process could help us find out how network structure, seed node, and information diffusion model might jointly affect information dissemination over social network.

11.4   Toolkit Illustrations

To study meme diffusion from a GIScience perspective, we adopt the concept of network communities as suggested by Weng et al. (2013) that demonstrate how clustered people form communities (in real space or virtual space) and form the entities in social networks that allow us to study, model and predict meme diffusion processes. Community structure has been demonstrated to affect information diffusion. It was also the basis for studying the activities of individuals and the speed of diffusion. Although the use of network communities as a concept to study meme diffusion is not new, we use agent-based simulations to model the spread of memes so that we not only model different forms of diffusion processes with simulations but also using simulated processes to enable prediction of ways meme diffusion may proceed over social networks. Agent-based modeling has been extensively used to predict human behavior like posting, forwarding or replying a message with regard to topics and sentiments (Ye and Lee 2016; Lee and Ye 2018).

We develop social media simulators using a bottom-up approach (micro level) to evaluate the emergent behavior (macro level). In this research, a network-based diffusion analysis (NBDA) has been developed to fit agent-based models of social and asocial learning to the observed data using maximum-likelihood estimation. The underlying learning mechanism is identified using model selection based on the Akaike information criterion. NBDA is able to discriminate between social and asocial learning in comparison with diffusion curve analysis. NBDA thus offers a more reliable statistical test of learning mechanisms. Spatiotemporal visualization is applied to intuitively reflect the complex process of information diffusion. These visualization techniques allow analysts to iteratively and interactively explore the dataset and thus gain deeper understanding of the origination, propagation, and clustering of information.

The main interface of this toolkit is composed of two parts, a menu bar on the top and a preview window below. Preview window holds a canvas for visualizing spatial data and network data. Menu bar contains eight drop-down lists that provide various groups of functions. File menu includes functions such as Open, Save, Save As and Close, which are designed for importing, manipulating, and exporting GIS and network files. Network menu is used to generate networks with different structures. Analysis menu provides functions to investigate attributes of nodes, edges, and networks. Community menu contains a group of algorithms regarding community detection. Text menu holds functions of content analysis and text data mining. Simulator menu contains the algorithms simulating information diffusion over spatial and virtual networks. Some of aforementioned menus can add a control panel to the main interface on the left of Preview such as Fig. 11.1. Windows menu is used to show or hide a specific control panel. Help menu provides information for users to check the background and tutorial of this tool.

After clicking Network menu, users can generate simulated social networks on the network generator control panel, and display the simulated networks on the map on the Preview Canvas (Fig. 11.1). If the background geographic map is needed, users can check Base Map on the control panel. Then users will be guided to click Select button to open shape file and click View button to display the shape file on the Canvas. Users can zoom out or zoom in base map on the Canvas view (Fig. 11.1) with mouse wheel. After configuring parameters of the selected network type, users can click Generate Network button to create the simulated network and display it on the top of the base map. It should be noted that the availability of parameters here depends on the selected network type. Edge checkbox is a switch to show or hide edges among nodes. Users can always delete the social networks on the base map by clicking Reset Network button on the control panel (Fig. 11.1).

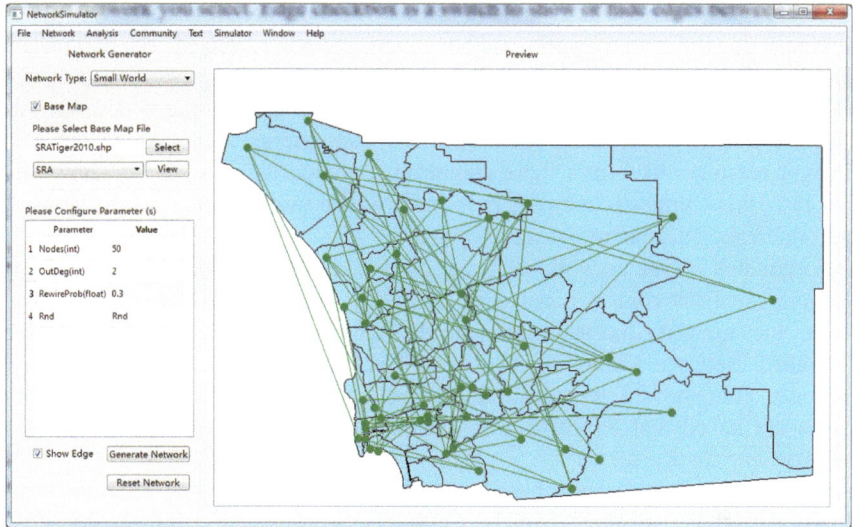

Fig. 11.1 Network generator

In addition to generating a single model network, the tool also supports user-generated composite networks (Fig. 11.2). The composite network can be developed by connecting multiple network models through selecting a group of nodes based on some predefined rules. These networks can belong to the same or

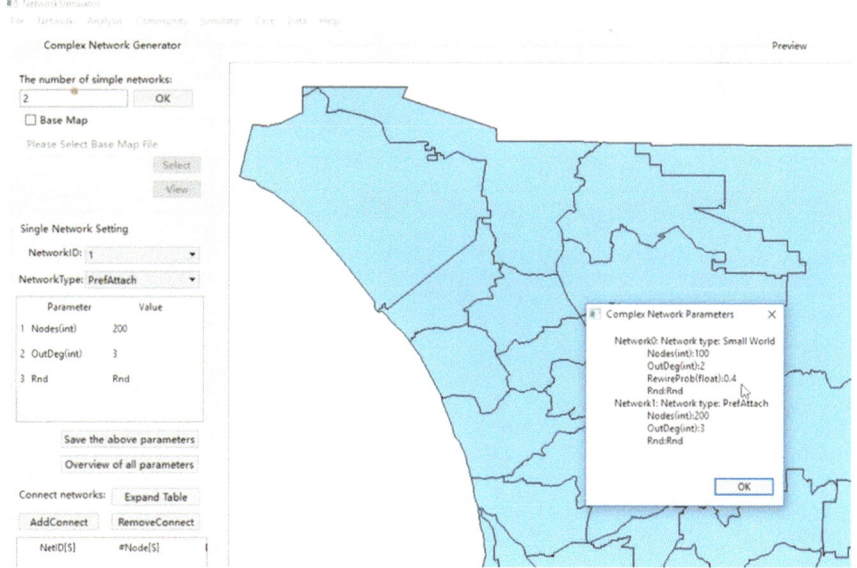

Fig. 11.2 Composite network

different model. The key to generating a composite network is how to select nodes for connecting different networks and how many nodes are selected. The parameters can be set by the researcher. Users can specify the number or proportion of nodes selected from a network, and then specify the rules to select these nodes, such as selecting a node by a certain Centrality and then adding a connection between different networks to form a more complicated composite network. As shown by Fig. 11.2, this composite network is formed by two models: small-world network and preference attachment network.

To conduct social network analysis, users can click Analysis menu to show Network Analysis control panel (Fig. 11.3). By clicking Run button, users can derive the network measurements in the table widget. By clicking Degree Centrality, Between Centrality, Closeness Centrality, or Eigenvector Centrality, users can compute different centralities measurements for each node. Centrality values are saved and exported with the name of centrality indicator, such as "DegreeCentralityResult.txt".

In the Community menu, there are three sub menus accessing different community detection algorithms, namely, CNM, Girvan Newman, and OSLOM. After users click a specific community detection sub menu, nodes are rendered by colors representing different communities (Fig. 11.4). At the same time, a result txt file will be created and exported based on this calculation as "results.txt", which stores nodes ids by community.

Text Analysis is a tool for content analysis on tweets (posts of twitter users). Text analysis helps reveal why some messages go viral from the semantic perspective. The integration of space, time, network, and content analysis facilitates the understanding of information diffusion over social network (Wang et al. 2016).

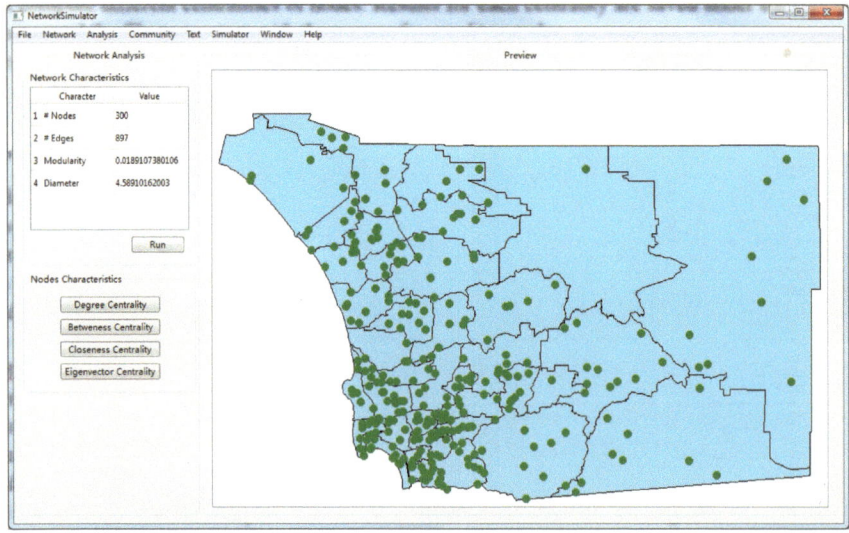

Fig. 11.3 Network analysis

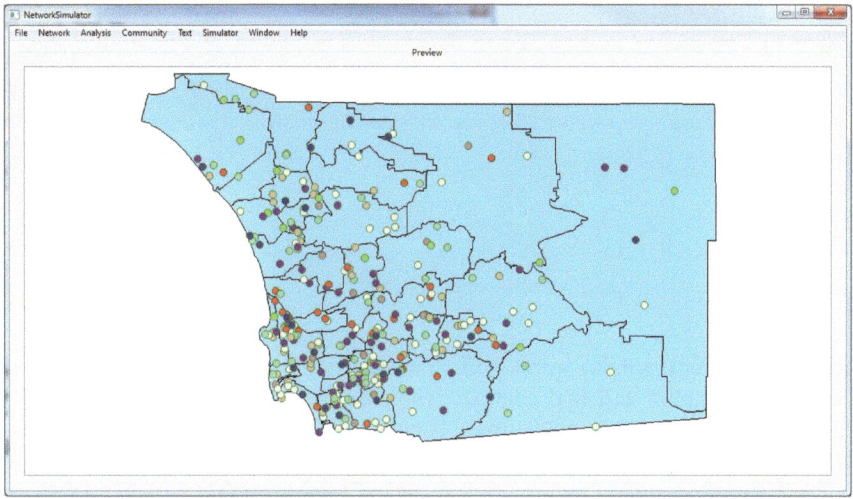

Fig. 11.4 Results of community detection

There are six functions in the Text Analysis tool (Fig. 11.5). Before running any of these functions except Frequency Distribution, users need to specify one or multiplekeywords: (1) clicking Search Text button will search all tweets; (2) clicking Search Similar Text button will identify other words based on the distributional

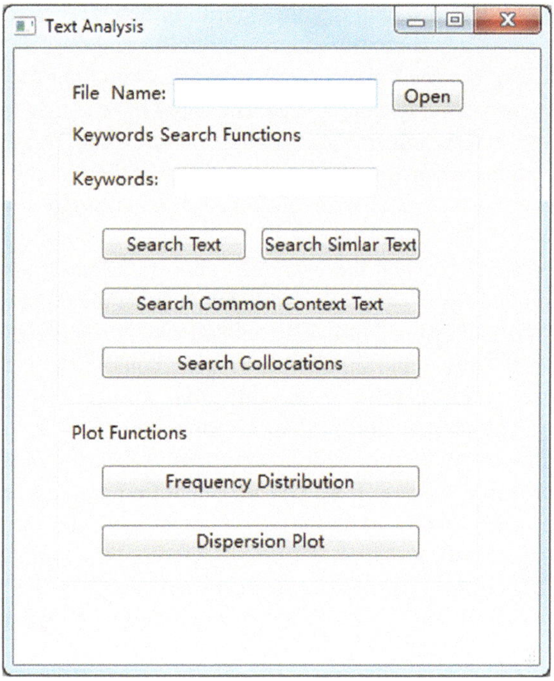

Fig. 11.5 Text analysis

Fig. 11.6 Results of
frequency distribution

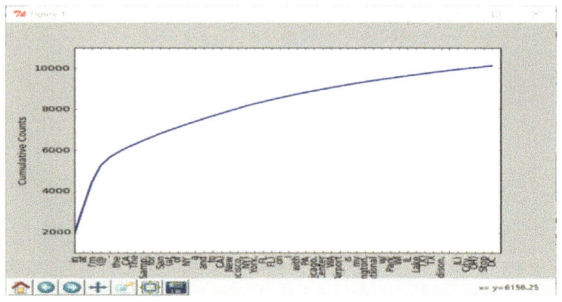

similarity and most similar words will be listed first; (3) clicking Search Common Context Text button will find contexts where the specified words appear and list most frequent common contexts first; (4) clicking Search Collocations button will derive the collocations from the text. A collocation is a sequence of words or terms that co-occur more often than would be expected by chance; (5) clicking Dispersion Plot button draws the location of the keyword in the text to visualize the changes of language use over time. Clicking Frequency Distribution button plots cumulative frequency of the most used words in the text (Fig. 11.6).

In addition to the analysis of historical tweets in the aforementioned functions, users can click Simulator menu to display Simulating Information Diffusion panel (Fig. 11.7). In this panel, users need to specify the number of seed nodes and a specific algorithm, then Click Generate Seed Node button to get blue nodes. In order to maximize the influence of information, there are seven algorithms: greedy algorithm, discount degree algorithm, betweenness centrality, closeness centrality, eigenvector centrality, degree centrality and random algorithm, which allow users to explore seed nodes (early adopters) initiating the spread of information. In the social network, opinion leaders usually have a greater authority and tend to be more active in the information sharing and diffusion. This toolkit allows users to specify a certain percentage of nodes as opinion leaders in the entire network or in a specific community. After selecting a propagation model, independent cascade model (IC model) or linear threshold model (LT model) can be chosen with specified parameters. Then the users can click Information Diffusion button to conduct the simulation. An animation will demonstrate the process of message propagating over the simulated network. The red nodes represent the nodes that have been influenced (Fig. 11.7). The simulation process could help the users find out how various driving forces affect meme diffusion.

The propagation probability of opinion leaders and normal nodes can be set differently to run various scenarios of information diffusion. The appropriate probability parameters can be identified through many times of iteration. The procedure is to set an initial parameter pair with the increment of each probability value until the final diffusion outcome reaching closer to the actual situation (Fig. 11.8). Locating optimal early adopters towards efficient information diffusion would depend on the network structure and propagation probabilities of nodes in

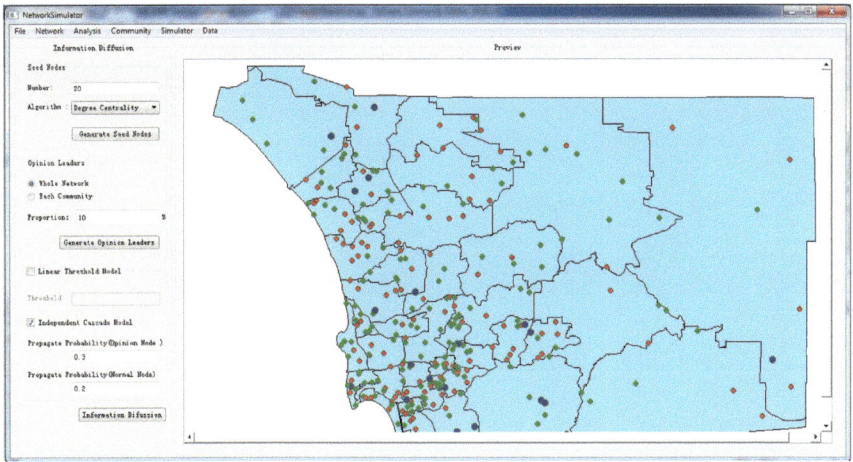

Fig. 11.7 Information diffusion

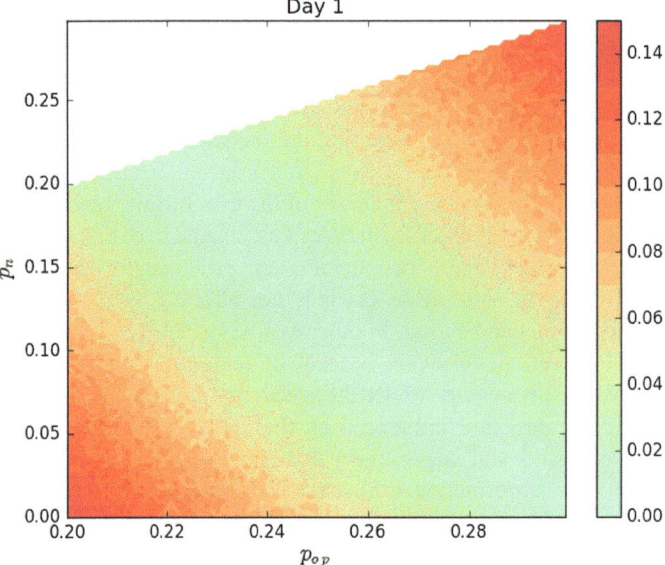

Fig. 11.8 Contour map of errors illustrating the information diffusion model on the propagation probabilities of opinion leaders (X) and normal nodes (Y)

the network. The contour map (Fig. 11.8) plots the areas of errors (difference between simulated results and real information diffusion outcomes). Areas of light green indicate areas with fewer errors.

The information diffusion model at the city level can also be designed to simulate how the occurrence of events spread in the urban network. The model assumes

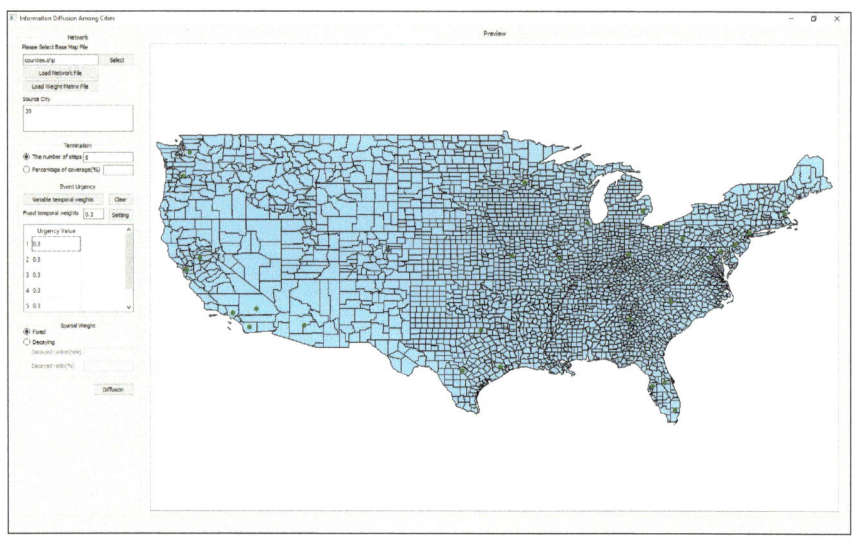

Fig. 11.9 Information diffusion across urban network

that there is an urban network with the nodes representing the cities and the edges between cities indicating the communication strength between cities. When a specific event occurs in one or more cities, the model simulates how the event has developed over a certain number of days (steps) and diffused to other cities. The model allows users to set the daily intensity of the event (Emergency Index), which records its ability to spread out. The intensity can change over time (Fig. 11.9). For the influence of spatial weight, two models are provided for the user to select (Fig. 11.10). Because a nationwide event is not affected by the distance among cities, the Fixed option can be used. In other words, spatial weight is not considered in this case. However, if the event is local or regional, the Decaying option is needed. That is, it is necessary to set the Decaying radius and Decaying ratio to reflect the diffusion structure influenced by the distance among cities. Given the same number of nodes and edges, the network with higher average path length tends to have wider information diffusion. The trend of simulation statistics is

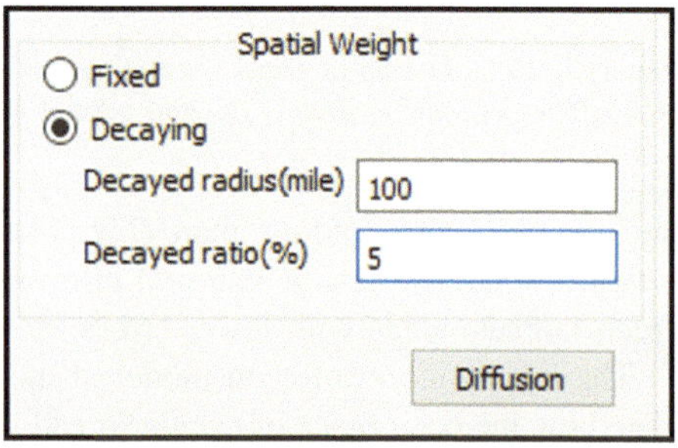

Fig. 11.10 Spatial weight in information diffusion

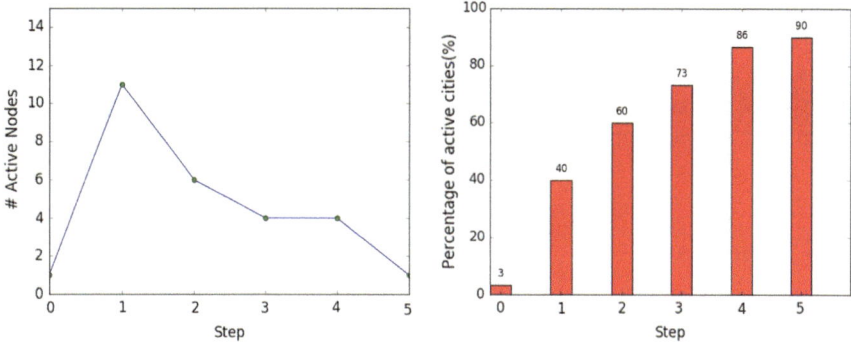

Fig. 11.11   Statistics of information diffusion across cities

presented in Fig. 11.11. The left figure reports the newly involved cities in each step of information diffusion while the right figure shows the cumulative percentage over step.

Figure 11.12 demonstrates a real-world case of wild fire taking place at the north of San Diego County from May 13 through May 23 in 2014 as an example. We collected all tweets within 40 miles around the center of San Diego during that time. The top right figure inside Fig. 11.12 in shows both the process of tweets posted in the space over time and the regression curve of the nearest neighborhood ratio for the purpose of estimating spatial diffusion processes (Lee et al. 2014). We use this

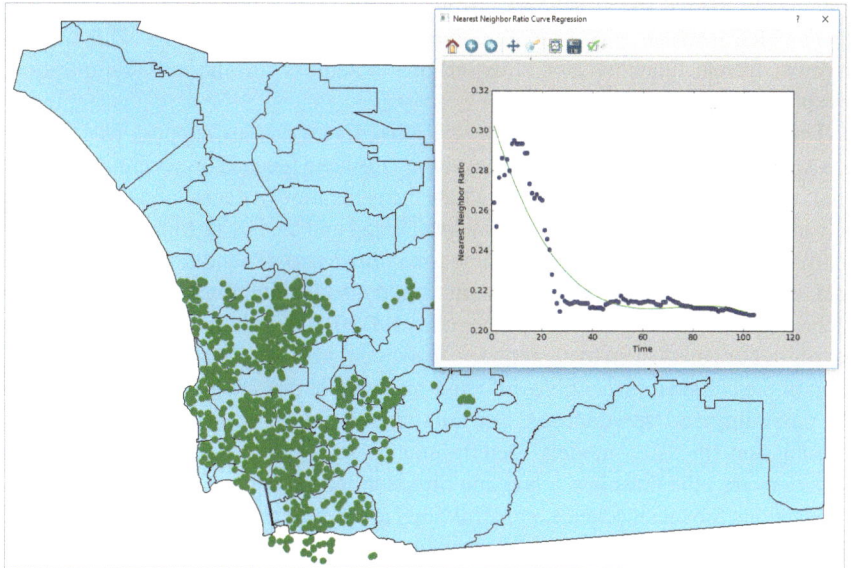

Fig. 11.12   A real world case of information diffusion

ratio to test the degree of spatial cluster of geo-tagged tweets, in order to model spatial diffusion processes. The main characteristics of information diffusion processes can then be detected and distinguished (Li et al. 2016). According to the process and the curve, users can better understand the pattern of message propagation over a social network in the real world.

11.5   Summary

As Sui (2014) noted, "(big) geospatial information and perspectives are becoming an imperative for major decisions at both individual and collective levels". Social media data such as tweets is fast becoming a source of geographic data, information, and perhaps even knowledge. There is a long tradition of linking relational and physical spaces for geographical issues across scales. Spatial and social network analyses also correspond to the notions of space and place, respectively. Moreover, beyond academia, government agencies, practitioners, as well as engaged citizens are increasingly recognizing the linkages between social space and physical environment. Rigorous analysis and simulation of meme diffusion is likely to open up a rich context for advancing our knowledge on complex socioeconomic events and bettering policy interventions. Interdisciplinary approaches of exploring and simulating rich and complex geo-tagged social media dynamics are highly demanded to ignite transformative geospatial-social innovation and discovery for enabling effective and timely solutions to challenging social and environmental problems (Wang et al. 2017). In this package, we develop models and algorithms to generate and analyze spatial social networks as well as simulate information diffusion. This open-source package can promote collaboration among researchers who want to improve current functions or add extensions to address specific research questions on spatial meme diffusion.

This toolkit has its origins as a specialized program to support research on geo-tagged tweets analysis. Recent developments have focused on several directions. First, we will develop a schema to facilitate the generation of data analysis tasks based on various dimensions of social media data (Wang and Ye 2017). More simultaneous analyses can be developed with the increasing dimensions. In addition to social media data, informative human-centric contents could also be found in many other sources such as OpenStreetMap and newspaper (Hu et al. 2017). Epidemiology-based models have also been used for investigating social network phenomenon. We will use a novel form of the classical susceptible–infected–susceptible disease model which includes the possibility for 'spontaneous' (or 'automatic') infection, in addition to disease transmission (the SISa model). Determining to what extent particular emotions or behaviors are infectious is a promising direction for further research with important implications for social science, epidemiology and health policy. The next-step ABM provides a theoretical framework for studying the interpersonal spread of any state that may also arise spontaneously, such as emotions, behaviors, health states, ideas or diseases with reservoirs. Due to the growing social media data size, a high performance computing power and related packages are needed to be implemented to facilitate the further growth of this toolkit. Based on the strength of interactive visualization techniques, this research stresses the need to study the geography of social network. On this basis, the sincere hope here is that this dialogue between social network analysis and GIS will embrace the real-world challenges and opportunities of big social data.

Acknowledgements This material is based upon work supported by the National Science Foundation under Grant No. 1416509 and 1637242.

References

Allaway, A. W., Berkowitz, D., & D'Souza, G. (2003). Spatial diffusion of a new loyalty program through a retail market. *Journal of Retailing, 79*(3), 137–151.

Allaway, A. W., Black, W. C., Richard, M. D., & Mason, J. B. (1994). Evolution of a retail market area: An event history model of spatial diffusion. *Economic Geography, 70*(1), 23–40.

Bailey, N. (1975). *The mathematical theory of infectious diseases and its applications.* London: Charles Griffin & Company Ltd.

Bakshy, E., Rosenn, I., Marlow, C., & Adamic, L. (2012, April). The role of social networks in information diffusion. In *Proceedings of the 21st International Conference on World Wide Web* (pp. 519–528). ACM.

Barabási, A. L., & Albert, R. (1999). Emergence of scaling in random networks. *Science, 286*(5439), 509–512.

Barrat, A., & Weigt, M. (2000). On the properties of small-world network models. *The European Physical Journal B-Condensed Matter and Complex Systems, 13*(3), 547–560.

Baybeck, B., & Huckfeldt, R. (2002). Urban contexts, spatially dispersed networks, and the diffusion of political information. *Political Geography, 21*(2), 195–220.

Budak, C., Agrawal, D., & Abbadi, A. E. (2011). Limiting the spread of misinformation in social networks. In *Proceedings of the 20th International Conference on World Wide Web (WWW '11)*.

Cao, N., et al. (2012). Whisper: tracing the spatiotemporal process of information diffusion in real time. *IEEE Transactions on Visualization and Computer Graphics, 18*(12), 2649–2658.

Centola, D. (2010). The spread of behavior in an online social network experiment. *Science, 329*(5996), 1194–1197.

Chesterman, A. (2016). *Memes of translation: The spread of ideas in translation theory* (Vol. 123). Amsterdam: John Benjamins Publishing Company.

Clauset, A., Newman, M. E., & Moore, C. (2004). Finding community structure in very large networks. *Physical Review E, 70*(6), 066111.

Doo, M. (2012). *Spatial and social diffusion of information and influence: Models and algorithms* (Doctoral dissertation, Georgia Institute of Technology).

Erdös, P., & Rényi, A. (1960). On the evolution of random graphs. *Publications of the Mathematical Institute of the Hungarian Academy of Sciences, 5*(17–61), 43.

Granovetter, M. S. (1987). Threshold models of collective behavior. *American Journal of Sociology, 83*(6), 1420–1443.

Girvan, M., & Newman, M. E. (2002). Community structure in social and biological networks. *Proceedings of the National Academy of Sciences, 99*(12), 7821–7826.

Goldenberg, J., Libai, B., & Muller, E. (2001). Talk of the network: A complex systems look at the underlying process of word-of-mouth. *Marketing Letters, 12*(3), 211–223.

Gregory, D., & Urry, J. (1985). Suspended animation: The stasis of diffusion theory. In D. Gregory & J. Urry (Eds.), *Social relations and spatial structures* (pp. 296–336). New York: St. Martin's Press.

Guille, A., & Hacid. H. (2012). A predictive model for the temporal dynamics of information diffusion in online social networks. In *WWW '12 Companion* (pp. 1145–1152).

Hägerstrand, T. (1967). Aspects of the spatial structure of social, communication and the diffusion of information. *Papers in Regional Science, 16*(1), 27–42.

Hanneman, R. E. (2000). *Introduction to social network methods. Online textbook supporting sociology 157.* Riverside, CA: University of California.

Hu, Y., Ye, X., & Shaw, S. L. (2017). Extracting and analyzing semantic relatedness between cities using news articles. *International Journal of Geographical Information Science, 31*(12), 2427–2451.

Kempe, D., Kleinberg, J., & Tardos, E. (2003). Maximizing the spread of influence through a social network. In *Proceedings of the ninth ACM SIGKDD International Conference on Knowledge Discovery and Data Mining (KDD '03).*

Knoke, D. (1982). The spread of municipal reform: Temporal, spatial, and social dynamics. *American Journal of Sociology, 87*(6), 1314–1339.

Lancichinetti, A., Radicchi, F., Ramasco, J. J., & Fortunato, S. (2011). Finding statistically significant communities in networks. *PLoS ONE, 6*(4), e18961.

Lee, J., Lay, J. G., Chin, W. C. B., Chi, Y. L., & Hsueh, Y. H. (2014). An experiment to model spatial diffusion process with nearest neighbor analysis and regression estimation. *International Journal of Applied Geospatial Research, 5*(1), 1–15.

Lee, J., & Ye, X. (2018). An open source spatiotemporal model for simulating obesity prevalence. In *GeoComputational Analysis and Modeling of Regional Systems* (pp. 395–410). Cham: Springer.

Liang, Y., Caverlee, J., Cheng, Z., & Kamath, K. Y. (2013). How big is the crowd? Event and location based population modeling in social media. In *Proceedings of the 24th ACM Conference on Hypertext and Hypermedia.* Paris, France.

Li, M., Ye, X., Zhang, S., Tang, X., & Shen, Z. (2017). A framework of comparative urban trajectory analysis. *Environment and Planning B.* https://doi.org/10.1177/2399808317710023.

Li, S., Ye, X., Lee, J., Gong, J., & Qin, C. (2016). Spatiotemporal analysis of housing prices in China: A big data perspective. *Applied Spatial Analysis and Policy.* https://doi.org/10.1007/s12061-016-9185-3.

Morrill, R., Gaile, G. L., & Thrall, G. I. (1988). Spatial diffusion. SAGE scientific geography series 10. Newbury Park, CA: SAGE Publications, Inc.

Newman, M. E. (2008). The mathematics of networks. *The New Palgrave Encyclopedia of Economics, 2*(2008), 1–12.

Newman, M. E. J. (2003). The structure and function of complex networks. *Society for Industrial and Applied Mathematics (SIAM) Review, 45*(2), 167–256.

Newman, M. E. J., Forrest, S., & Balthrop, J. (2002). Email networks and the spread of computer viruses. *Physical Review, 66*(3), 035101.

Romero, D. M., Tan. C., & Ugander, J. (2013). On the interplay between social and topical structure. In *Proceedings of AAAI International Conference on Weblogs and Social Media* (pp. 516–525).

Romero, D. M., Galuba, W., Asur, S., & Huberman, B. A. (2011, March). Influence and passivity in social media. In *Proceedings of the 20th international conference companion on World wide web* (pp. 18–33). ACM.

Scott, J. (2012). *Social network analysis.* UK: Sage.

Shaw, S., Tsou, M., & Ye, X. (2016). Human dynamics in the mobile and big data era. *International Journal of Geographical Information Science, 30*(9), 1687–1693.

Spitzberg, B. H. (2014). Toward a model of meme diffusion (M3D). *Communication Theory, 24*(3), 311–339.

Sui, D. (2014). Opportunities and impediments for open GIS. *Transactions in GIS, 18*(1), 1–24.

Wang, F., Wang, H., & Xu, K. (2012). Diffusive logistic model towards predicting information diffusion in online social networks. In *ICDCS '12 Workshops* (pp. 133–139).

Wang, S. (2016). CyberGIS and spatial data science. *GeoJournal.* https://doi.org/10.1007/s10708-016-9740-0.

Wang, Y. D., Fu, X. K., Jiang, W., Wang, T., Tsou, M. H., & Ye, X. (2017). Inferring urban air quality based on social media. *Computers, Environment and Urban Systems, 66,* 110–116.

Wang, Z., & Ye, X. (2017). Social media analytics for natural disaster management. *International Journal of Geographical Information Science.* https://doi.org/10.1080/13658816.2017.1367003.

Wang, Z., Ye, X., & Tsou, M. (2016). Spatial, temporal, and content analysis of Twitter for wildfire hazards. *Natural Hazards.* https://doi.org/10.1007/s11069-016-2329-6.

Watts, D. J., & Strogatz, S. H. (1998). Collective dynamics of 'small-world' networks. *Nature, 393* (6684), 440–442.

Weng, L., Menczer, F., & Ahn, Y.-Y. (2013). Virality prediction and community structure in social networks. *Scientific Reports, 3,* 2522.

Yang, J., & Counts, S. (2010). Predicting the speed, scale, and range of information diffusion in Twitter. In *4th International AAAI Conference on Weblogs and Social Media (ICWSM).*

Yang, J., & Leskovec, J. (2010). Modeling information diffusion in implicit networks. In *ICDM '10: IEEE International Conference On Data Mining, 2010.*

Ye, X. (2017). Open data and open source GIS, In Huang, B. (Ed.), *Comprehensive geographic information systems* (Vol. 1, pp. 42–49). Oxford: Elsevier. https://doi.org/10.1016/b978-0-12-409548-9.09592-0.

Ye, X., Li, S., Yang, X., Lee, J., & Wu, L. (2018). The fear of Ebola: A tale of two cities in China. In *Big data support of urban planning and management* (pp. 113–132). Cham: Springer.

Ye, X., & Lee, J. (2016). Integrating geographic activity space and social network space to promote healthy lifestyles. *ACMSIGSPATIAL Health GIS., 8*(1), 24–33.

Ye, X., & Rey, S. J. (2013). A framework for exploratory space-time analysis of economic data. *Annals of Regional Science, 50*(1), 315–339.

Zhang, J., Liu, X., & Tong, Z. (2012). Natural disaster risk assessment using information diffusion and geographical information system. In: Lu, J., Jain, L., & Zhang, G. (Eds.), *Handbook on decision making* (pp. 309–330). Berlin, Heidelberg: Springer.

Author Biographies

Xinyue Ye (Ph.D., University of California at Santa Barbara and San Diego State University), is an associate professor in the Department of Geography at Kent State University where he directs computational social science lab. His major expertise is on modelling the geographical perspective of socioeconomic inequality and human dynamics. He develops and implements new methods on spatiotemporal-social network analysis/modelling/simulation for different application domains such as economic development, disaster response, land use, public health, and urban crime. He has received about 4.5 million dollars in grants as PI or Co-PI from National Science Foundation, Department of Commerce, and Department of Energy.

Lanxue Dang (Ph.D., Henan University), is an associate professor in the College of Computer and Information Engineering, Henan University. He is interested in spatial meme diffusion and open source toolkit development. He has also studied spatial optimization and bus routing issues.

Jay Lee (Ph.D., University of Western Ontario), is a professor in the Department of Geography at Kent State University. His interests include relating geographic events and patterns through time and space. This includes quantitatively modeling changing geographic events as a diffusing spatial processes. This may be applied to many aspects of our daily life: land use, air pollution, crime, public health, and how all these interact with environment. Some of my publications and research grants have involved digital elevation models, environmental conservation, GIS, web-based GIS, urban growth, urban sprawl, management of urban growth and areal health disparities.

Ming-Hsiang Tsou (Ph.D., University of Colorado, Boulder), Professor, Department of Geography, San Diego State University; Founding Director, Center for Human Dynamics in the Mobile Age (HDMA) (http://humandynamics.sdsu.edu). His research interests are in Big Data, Human Dynamics, Social Media, Visualization, and Internet GIS. He is co-author of Internet GIS, a scholarly book published in 2003 by Wiley and served on the editorial boards of the Annals of GIS (2008-), Cartography and GIScience (2013-) and the Professional Geographers (2011-). His recent NSF research projects focus on health disparities, disaster responses, and spatial social networks.

Zhuo Chen (M.A., Kent State University), is a doctoral candidate of GIS in the Department of Geography at Kent State University. He has been widely trained in spatial programming, transportation analysis, and open source geocomputation. He has the research expertise on agent based modelling of meme diffusion.

Chapter 12
The Opportunities and Challenges with Social Media and Big Data for Research in Human Dynamics

Atsushi Nara, Ming-Hsiang Tsou, Jiue-An Yang and Cheng-Chia Huang

12.1 Introduction

Social media and Big Data have transformed our daily lives into interconnected cyberspace and realspace (Shaw and Yu 2009; Tsou 2015). As more location-aware technologies becoming available, social media platforms have increasingly embraced the location-based dimension (Sui and Goodchild 2011) and GIScience have attracted more interest in the dynamic relations of human behaviors and the environment (Shaw et al. 2016). Geographers can now collect, trace, and visualize the spread of social movements, disease outbreaks, nature hazards, and popular events by digitally collecting social media and Big Data with locational contents (Tsou 2015). This is largely due to the advances in location sensing and information and communication technologies, in particular on a mobile platform. These new technologies enable automatic tracking of human movement and behavior outdoor and indoor at a high level of details in space and time using location-aware technologies (LATs) such as global positioning systems (GPS), cellular networks, WiFi positioning system, Radio-Frequency Identification Device (RFID), surveillance camera, and various kinds of portable smart devices with LATs. Spatially and temporally fine-granular timestamped location data can reconstruct individual trajectories and describe dynamic movement behaviors in detail. In addition, individual-scale data can avoid conventional data scaling problems such as ecological fallacy and modifiable areal unit problems by aggregating data from bottom-up, allowing researchers examining both individual and collective behavior. Furthermore, the pervasiveness of smartphone and internet usage as well as the

A. Nara (✉) · M.-H. Tsou · C.-C. Huang
Department of Geography, San Diego State University, San Diego, CA, USA
e-mail: anara@mail.sdsu.edu

J.-A. Yang
Qualcomm Institute, University of California, La Jolla, San Diego, CA, USA

© Springer International Publishing AG, part of Springer Nature 2018
S.-L. Shaw and D. Sui (eds.), *Human Dynamics Research in Smart
and Connected Communities*, Human Dynamics in Smart Cities,
https://doi.org/10.1007/978-3-319-73247-3_12

increasing trend of social media usage accelerate the generation of social media and big data with location information. The dynamic characteristics of social media and Big Data offer geographers research opportunities for examining and modeling human behaviors, communications, and movements (Tsou 2015). This short viewpoint paper reports on a summary of papers presented in a series of special sessions, Human Dynamics in the Mobile Age: Linking Physical and Virtual Spaces, at the Association of American Geographers (AAG) annual meeting in 2015 and Symposium on Human Dynamics Research: Social Media and Big Data at the AAG annual meeting in 2016. The summative report is categorized into three research components in these papers: data, method, and application. In addition, we discuss the current state-of-the-arts in human dynamics research and highlight their key concepts, opportunities, and challenges.

12.2 Human Dynamics Research: Summary of Papers in AAG Special Sessions

Human dynamics is a transdisciplinary research field focusing on the understanding of dynamic patterns, relationships, narratives, changes, and transitions of human activities, behaviors, and communications. The advent of location aware technologies, ubiquitous network infrastructures, and mobile technologies accelerate human dynamics research by providing opportunities for researchers to access to a large amount of fine-granular individual-scale data, which were not available in the past. The availability of such social media and big data is leading to a data-driven scientific inquiry, which is purely inductive and emergent forms of analysis that data to speak for itself (Kitchin 2014; Kwan 2016). To encourage more geographers and GIScientists to study this emerging research themes, a series of special sessions were organized at AAG annual meetings in 2015 and 2016, *Human Dynamics in the Mobile Age: Linking Physical and Virtual Spaces* (6 sessions) and *Symposium on Human Dynamics Research: Social Media and Big Data* (3 sessions) respectively, and the total of 42 papers were presented. We analyzed paper titles, abstracts, and keywords and summarized this new research theme by analyzing the data, methods, and applications from the 42 representative research abstracts (Table 12.1).

12.2.1 Data

Of the 42 papers presented at the AAG human dynamics sessions, just over half (n = 22, 52.4%) used social media data, which was broken down into Facebook (n = 1), Flickr (n = 3), Foursquare (n = 1), Instagram (n = 2), Twitter (n = 16), and Weibo (n = 1). These counts are not mutually exclusive and one paper used data from multiple social media. The social media data can be gathered via Application Programming Interfaces (APIs), which allow users to access to publicly

Table 12.1 Characteristics of papers presented at the AAG annual meetings in 2015 and 2016

Total number of papers			2015	2016	Total
			(n)	(n)	(n)
			27	15	42
Data	Activity record		1		1
	Cadastral record		1		1
	Camera/video image		2		2
	Census			1	1
	Interview/survey		1		1
	Mobile phone data (CDRs/SMS)		7	1	8
	Social media		12	10	22
	GPS tracks		1		1
	Volunteer geographic information			2	2
	Unknown/no data		2	1	3
Method[a]	GIS	GIS/Web-GIS (general)	4	6	10
		Spatial analysis	2		2
		Spatial statistics	1		1
		Spatiotemporal analysis	7	2	9
		Visualization	2	1	3
	Modeling	Geosimulation	1		1
		Spatial interaction	1		1
		Statistical modeling/analysis	3	2	5
	Data mining/machine learning		4	3	7
	Literature review/overview		1	3	4
	Network/graph analysis		2		2
	Qualitative method		1	1	2
	Text mining/semantic analysis		6	3	9
	Trajectory analysis		2		2
Application[a,b]	Disaster/risk		3	2	5
	GIS/Web-GIS tools		2		2
	Health		3	1	4
	Human mobility/movement behavior		9	6	14
	Marketing			1	1
	Communication		1	1	2
	Realspace-cyberspace		3	2	5
	Transportation		3		3
	Urban dynamics		2	2	4

[a]These counts are not mutually exclusive
[b]Several studies do not mention any specific application

available social media contents, or purchased from social media providers. Types of social media data that researcher can have access vary by social media platforms. These include, for example, media contents such as text messages, photos, and check-ins, tags, timestamps, and locations on media contents, user profile, and user's social network relationships. As smartphones have become pervasive in everyday life, location attributes are often associated with users' mobile phone location acquired by GPS, cellular networks, or assisted GPS (A-GPS) supported by cellular networks. Data used in the remaining papers include mobile phone data such as Cell Detail Records (CDRs) and Short Message Service (SMS) (n = 8), camera/video imagery (n = 2), Volunteer Geographic Information (VGI) (n = 2), GPS (Global Positioning System) tracks (n = 1), US Census (n = 1), activity record (n = 1), cadastral record (n = 1), and interview/survey (n = 1).

12.2.2 Methods

Among a variety of methodological approaches were presented, the largest mentioned was GIS (n = 25) as a general framework and a tool to analyze and visualize human dynamics data in conjunction with other analytical methodologies. Under the GIS category, 10 papers mentioned GIS or Web-GIS as a base framework. Spatiotemporal analysis (n = 9) was the second largest GIS method mentioned in papers to study human dynamics in both spatial and temporal dimensions and most papers mentioned it in conjunction with other specific methodologies such as trajectory analysis and text mining. Since social media data often contain text data, text mining and semantic analysis appeared as a popular analytical methodology (n = 9). Specific methodologies applied to text data include, for example, Support Vector Machine (SVM) and Latent Dirichlet Allocation (LDA) for finding text similarities and topics. Papers mentioned data mining and machine learning as general data analytics frameworks (n = 7), while others employed social network/graph-based analysis (n = 2) and trajectory analysis (n = 2). Spatial modeling approaches were presented to describe process and flow of human dynamics and their behavior including geosimulation (n = 1) and spatial interaction (n = 1). Other methods included spatial statistics (n = 1), spatial analysis (n = 2), statistical analysis/modeling (n = 5), visualization (n = 3), literature reviews (n = 4), participant observation/interview as a qualitative method (n = 2), and an overview discussing challenges and opportunities of human dynamics research (n = 1).

12.2.3 Applications

A total of 15 papers applied to study general human dynamics and movement behavior including inter- and intra-urban population flows, tourist's movement, and human activity space. Studies also utilized social media and big data to examine human mobility, human behavior, and information flow in application to risk assessment and management during disastrous events (n = 5) (e.g., disaster alerts and responses), public health (n = 4) (e.g., infectious disease dynamics, diet behavior), urban dynamics (n = 5) (e.g., gentrification), transportation (n = 3) (e.g., driving and parking behavior), communication (n = 2) (e.g., public perception and information diffusion), and marketing (n = 1) (e.g., cyberspace interaction and consumer behavior). In addition, 5 papers examined human behaviors, communications, and movements and these relationships between cyberspace and realspace.

12.3 The Current State of the Arts in Human Dynamics Research

Papers presented at the AAG sessions covered a broad range of the current state of the arts research topics related to human dynamic research utilizing social media and big data, and related works have been reported in the recently published literature. In terms of data, more disaggregated geo-referenced social media and big data collected via LATs as well as conventional methods (e.g., Census survey) have been utilized to study human dynamics. A few examples are Instagram and Twitter to analyze urban dynamic activity and demographic patterns (Boy and Uitermark 2016; Longley et al. 2015), CDRs to assess the validity of using CDR data for understanding human mobility (Zhao et al. 2016), GPS and accelerometer data to examine physical activity related to built environments (Miller et al. 2015), and the Longitudinal Employer-Household Dynamics (LEHD) data to study disaggregated work trip flows socio-spatial interaction (Niedzielski et al. 2015). In addition, new web and mobile tools have been developed to effectively collect and analyze such social media and big data for human dynamics research (Yang et al. 2016). Furthermore, High Performance Computing (HPC) enables to simulate large-scale human dynamics where millions of agents move and interact in a virtual space under the framework of Agent-Based Modeling (ABM). Such geosimulation frameworks can generate massive microscopic human movement data for exploring and investigating complex streetscape dynamics (Torrens 2016).

Quite a few methodologies have been proposed to conduct research on human mobility, their behavior, and contexts at both disaggregated and aggregated scales. For example, human movement behavior and mobility contexts can be analyzed by examining statistical and geometric properties of human dynamics data (Dodge et al. 2012; Torrens et al. 2012). Space-time analytics can examine reoccurring movements of individuals and from the reoccurring movements to identify patterns of life and their opportunities for interactions based on proximity in space and time (Yuan and Nara 2015). The trajectory-based analysis is used to extract movement characteristics of surgical staff from data collected by an ultrasonic-based location aware system as well as video imagery, which can ultimately describe surgical contexts (Nara et al. 2017). Location-based social network attempts to find human interactions and community structures by creating and analyzing graph networks based on spatial and spatio-temporal constraints under the Time-geography framework (Crooks et al. 2016; Yuan et al. 2014). ABM simulates mobility, decision making process, human-human interaction, and human-environment interaction for modeling complex human dynamics over space and time (An et al. 2014; Heppenstall et al. 2012; Torrens 2015). Text mining and machine learning techniques can be applied to social media and big data to reduce noises and extract meaningful contexts (Allen et al. 2016).

Application examples include public health and epidemiology surveillance (Nagel et al. 2013), criminology (Malleson and Andresen 2015), social movements (Tsou et al. 2013), risk assessment and management for nature hazards and disastrous events (De Longueville et al. 2009; Wang et al. 2016), to name a few.

12.4   Research Opportunities

There are numerous research directions that researchers can take to investigate human dynamics utilizing social media and big data in the coming years. Here we present three examples, location-based social network, location-based linguistic analysis, and dynamic spatial ontology. The first research direction is along the line of the brining the spatial dimension to social network analysis (SNA) and integrating social networks (SNs) into GIS. Social networks are built on the basis of node-edge graph structures where the distance between nodes is the geodesic distance, i.e. the shortest path between two nodes. This distance is known as degree and fits well for modeling the relationships and influences in graphs. However, it ignores the fact that human activities happen at a specific location in physical space and the importance of physical distance is not considered in SNA. By bringing the spatial dimension to SAN, researchers can examine the spatial context and geometry alongside with the graph characteristics (Brockmann and Helbing 2013; Doreian and Conti 2012; Hristova et al. 2016). Nevertheless, there have been few attempts in developing metrics that can combined the existing SNA with spatial analysis to quantify interactions among nodes in the spatial context. In terms of the convergence of SNs and GIS, challenges remain when representing complex multilevel SNs in GIS at the conceptual level (Sui and Goodchild 2011). At the application end, some efforts can be seen in laying out guidelines for modeling various types of SNs in geographic space for understanding human behavior (Yuan and Nara 2015; Andris 2016). One application example is combining SNs and geovisual analytics to representations the spatially embedded SNs from social media. As shown in Fig. 12.1, adding spatial attributes to the social network from Twitter conversations can suggest how location and urban hierarchy might have impacts on how metropolitan areas response to information.

Location-based linguistic analysis is another promising research direction, which utilizes text mining techniques to study human dynamics related to feelings, emotions, and opinions about places extracted from a large amount of textual contents of georeferenced social media and big data. For example, sentiment analysis, a text classification method, can be used to investigate how geographic

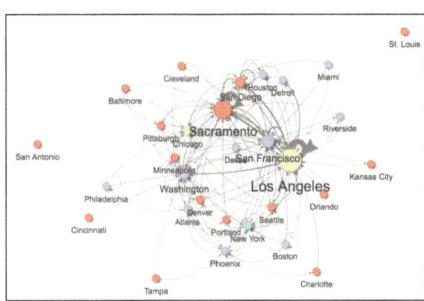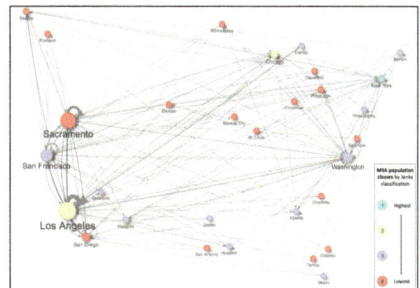

Fig. 12.1 An example of a spatial social network between the top 30 U.S. populated Metropolitan Statistical Areas (right) as compared to a regular social network (left). Directed edges between nodes indicate the frequencies of retweeting activities among different SRAs related to the California vaccine exemption conversations

places correlate with certain textual contents such as the levels of happiness (Mitchell et al. 2013). Topic modeling such as Latent Dirichlet Allocation (LDA) and Probabilistic Latent Semantic Indexing (PLSI) (Aggarwal and Zhai 2012) allows to explore the spatial patterns of themes discovered from geo-referenced text data; for example, the spatial patterns of health behavior topics like "childhood obesity and schools," "obesity prevention," and "obesity and food habits" (Ghosh and Guha 2013) and those of common topics on Twitter and their associations with demographic and socio-economic characteristics of Twitter users as well as places and local activities (Lansley and Longley 2016). Text clustering can be applied to group similar unstructured text documents into clusters and allows to investigate spatial clusters associated with built environments and place characteristics. For example, text documents including email correspondence, transcribed face-to-face interviews, and phone calls can provide new and important clues in a criminal investigation (Helbich et al. 2013).

Social media and big data also provide research opportunity to establish dynamic ontology for places (location names) and geographic regions. Traditional spatial ontology is defined by experts or gazetteer dictionaries which are difficult to formalized and standardized. We can define "place" ontology by aggregating hundreds of thousands of geo-tagged social media data (e.g., tweets) mentioned a specific place name (such as "SDSU") with linguistic analysis (Fig. 12.2). Cartographic visualization methods (e.g., kernel density estimate) can be used to identify the spatial boundary of place names, whereas content analysis can be employed to reveal the meaning of places. These methods can be further applied to observe the temporal changes of the boundaries associated with place names between different seasons. Different from the traditional definition of place names from gazetteers or experts, this new social media and big data-based ontology framework is human-centered and can provide useful information and operational meanings for places.

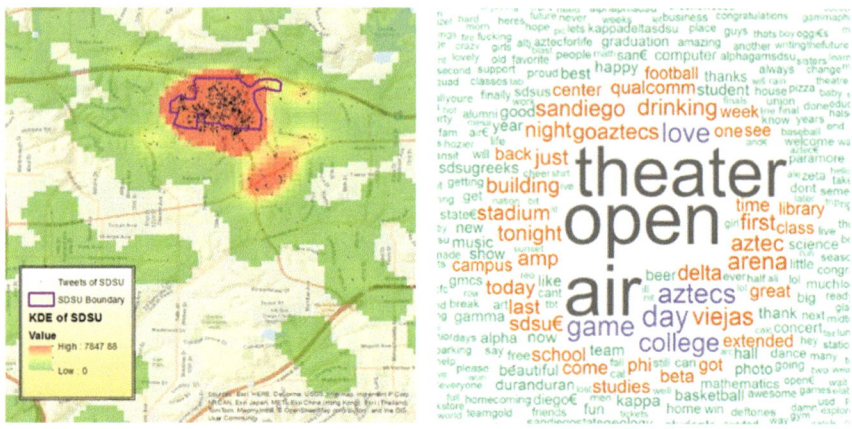

Fig. 12.2 The kernel density estimate hotspot of geotagged tweets containing the "SDSU" keyword (left) and the word could from the SDSU geotagged tweets (right)

12.5 Research Challenges

While social media data and big data provide new research opportunities, there exist notable challenges. Tsou (2015) listed seven research challenges related to mapping social media and big data; (1) lack of demographic profile, (2) data integration problems, (3) issues with user privacy and locational privacy, (4) needs of multi-disciplinary collaborations, (5) needs of contextual analysis, (6) filtering noises, and (7) difficulty of the falsifiability of hypotheses and theories. In addition to these challenges, we further identified two key challenges. One challenge relates to the fact that social media platforms/services and the internet of things (IOT) are dynamically evolving over time. APIs of a social media platform will be updated and major revisions on the service and the data access policy can affect data collection and possibly lead a data inconsistency issue. At the time of writing (October, 2016), APIs have been changed, for example, 14 times for Instagram since April 2014, 36 times for Flickr API since November 2010, 68 times for Foursquare API since November 2010, and 113 times for Twitter API since December 2012. Some involve major API changes; for instance, Twitter was originally designed as a text messaging service with a limitation of 160 characters including 20 characters for a user name and 140 characters for a message post. It is now allowing users to post a text message with emoji, images, and videos. These changes not only alter the data structure but likely influence user behavior, which makes human dynamics research using Twitter data more complex. Instagram has also made significant changes to its API in June 2016 that include the deprecation real-time subscriptions for tags, locations and geographies (an equivalent to the Twitter streaming API) and the mandatory requirement of a valid access-token to use APIs in order to fully access Instagram contents (Instagram 2016). To obtain a valid access-token, it requires researchers to develop a live application that has to be reviewed and approved by Instagram. The availability of social media data in a currently accessible data format, therefore, will likely be changed in the near future, which makes researchers especially difficult to conduct a longitudinal study.

Another key challenge is related to data and algorithm uncertainty. In spite of the emerging new research opportunities to produce geographic knowledge by utilizing social media and big data, most of these data are not the output of instruments designed to produce valid and reliable data amenable for scientific analysis (Lazer et al. 2014). Regarding the spatial data quality, location information in most social media and big data can be controlled by end users and it is challenging to know the level of uncertainty by researchers. For example, a location of an Instagram post is selected by a user based on a list of locations provided by Instagram; therefore, a user can easily manipulate his/her location. Furthermore, there exist quite a few web tools and mobile applications to fake location information. While these users' decision to fake location, or spoof location, protect individual's geo-privacy, few studies have discussed and incorporated location spoofing in the existing GIScience literature (Zhao and Sui 2017).

Kwan (2016) also questioned that big data-driven research ignores the potentially significant influence of algorithms on research results, and thus geographic knowledge generated with big data might be more of an artifact of the algorithms used than the data itself. For example, Fischer (Fischer 2014) mapped six billion geo-tagged tweets and observed a banding phenomenon, where the original tweet locations tend to align with the closest latitude or longitude, suggesting that tweet locations might have been fuzzed by Twitter through snapping them to the closest latitude or longitude to prevent people's exact locations being disclosed. Researchers often do not have access to, or even do not know about such algorithms being used by social media providers who generate, process, and provide their data through APIs. Moreover, in order to deal with big data, algorithms are increasingly implemented as computerized procedures, and they become increasingly detached from and less visible to researchers who use them (Kwan 2016). Consequently, such algorithms introduce greater uncertainty and potentially result in significant differences in research findings. Hence, it is crucial to examine and evaluate the validity of data and algorithms in order for maximizing the utility of social media and big data.

Addressing these 9 challenges will be an ongoing endeavor to move forward with human dynamics research utilizing new technologies, social media data, and big geospatial data. In two AAG special sessions, a few papers undertook some of these challenges. For example, two papers integrated more than two data sources; one utilized mobile phone location data, CDRs, and subway smartcard data to uncover dynamic urban population flow patterns, and the other combined Twitter, Flickr, and Instagram data to delineate dynamic place boundaries. One paper applied interview data to explore the practices, potentials, and problems in using data produced through mobile communications for disease disaster management.

While majority of papers presented in the AAG sessions focused on the exploratory data analysis revealing interesting patterns related to human dynamics, there is a need for human dynamics research tacking those challenges to critically discuss the use of new forms of data.

Acknowledgements This material is based upon work supported by the National Science Foundation under Grant No. 1634641, IMEE project titled "Integrated Stage-Based Evacuation with Social Perception Analysis and Dynamic Population Estimation" and Grant No. 1416509, IBSS project titled "Spatiotemporal Modeling of Human Dynamics Across Social Media and Social Networks". Any opinions, findings, and conclusions or recommendations expressed in this material are those of the author and do not necessarily reflect the views of the National Science Foundation.

References

Aggarwal, C. C., & Zhai, C. (2012). A survey of text clustering algorithms. In C. C. Aggarwal & C. Zhai (Eds.), *Mining text data* (pp. 77–128). Boston, MA: Springer.

An, L., Zvoleff, A., Liu, J., & Axinn, W. (2014). Agent-Based modeling in Coupled Human and Natural Systems (CHANS): Lessons from a comparative analysis. *Annals of the Association of American Geographers, 104,* 723–745.

Andris, C. (2016). Integrating social network data into GISystems. *International Journal of Geographical Information Science, 30,* 2009–2031.

Allen, C., Tsou, M.-H., Aslam, A., Nagel, A., & Gawron, J.-M. (2016). Applying GIS and Machine learning methods to Twitter data for multiscale surveillance of influenza. *PLoS ONE, 11,* e0157734.

Boy, J. D., & Uitermark, J. (2016). How to study the city on Instagram. *PLoS ONE, 11,* e0158161.

Brockmann, D., & Helbing, D. (2013). The hidden geometry of complex, network-driven contagion phenomena. *Science, 342,* 1337–1342.

Crooks, A. T., Croitoru, A., Jenkins, A., Mahabir, R., Agouris, P., & Stefanidis, A. (2016). User-generated big data and urban morphology. *Built Environment, 42,* 396–414.

De Longueville, B., Smith, R. S., & Luraschi, G. (2009). "OMG, from Here, I Can See the Flames!": A use case of mining location based social networks to acquire spatio-temporal data on forest fires. In *Proceedings of 2009 International Workshop Location Based Social Networks ACM* (pp. 73–80). New York, NY, USA.

Dodge, S., Laube, P., & Weibel, R. (2012). Movement similarity assessment using symbolic representation of trajectories. *International Journal of Geographical Information Science, 26,* 1563–1588.

Doreian, P., & Conti, N. (2012). Social context, spatial structure and social network structure. *Social Networks, 34,* 32–46.

Fischer, E. (2014). Making the most detailed tweet map ever. In: Mapbox. https://www.mapbox.com/blog/twitter-map-every-tweet/. Accessed 15 October 2016.

Ghosh, D., & Guha, R. (2013). What are we "tweeting" about obesity? Mapping tweets with topic modeling and Geographic Information System. *Cartography and Geographic Information Science, 40,* 90–102.

Helbich, M., Hagenauer, J., Leitner, M., & Edwards, R. (2013). Exploration of unstructured narrative crime reports: An unsupervised neural network and point pattern analysis approach. *Cartography and Geographic Information Science, 40,* 326–336.

Heppenstall, A. J., Crooks, A. T., See, L. M., & Batty, M. (Eds.). (2012). *Agent-Based models of geographical systems.* Netherlands, Dordrecht: Springer.

Hristova, D., Williams, M. J., Musolesi, M., Panzarasa, P., & Mascolo, C. (2016). Measuring urban social diversity using interconnected geo-social networks. In: *Proceedings 25th International Conference on World Wide Web. International World Wide Web Conferences Steering Committee* (pp. 21–30). Republic and Canton of Geneva, Switzerland.

Instagram. (2016). Instagram developer API. https://www.instagram.com/developer/. Accessed 23 August 2016.

Kitchin, R. (2014). Big data, new epistemologies and paradigm shifts. *Big Data & Society, 1,* 2053951714528481.

Kwan, M.-P. (2016). Algorithmic geographies: Big data, algorithmic uncertainty, and the production of geographic knowledge. *American Association of Geographers Annals, 106,* 274–282.

Lansley, G., & Longley, P. A. (2016). The geography of Twitter topics in London. *Computers, Environment and Urban Systems, 58,* 85–96.

Lazer, D., Kennedy, R., King, G., & Vespignani, A. (2014). The parable of Google flu: Traps in big data analysis. *Science, 343,* 1203–1205.

Longley, P. A., Adnan, M., & Lansley, G. (2015). The geotemporal demographics of Twitter usage. *Environment and Planning, 47,* 465–484.

Malleson, N., & Andresen, M. A. (2015). The impact of using social media data in crime rate calculations: Shifting hot spots and changing spatial patterns. *Cartography and Geographic Information Science, 42,* 112–121.

Miller, H. J., Tribby, C. P., Brown, B. B., Smith, K. R., Werner, C. M., Wolf, J., et al. (2015). Public transit generates new physical activity: Evidence from individual GPS and accelerometer data before and after light rail construction in a neighborhood of Salt Lake City, Utah, USA. *Health Place, 36,* 8–17.

Mitchell, L., Frank, M. R., Harris, K. D., Dodds, P. S., & Danforth, C. M. (2013). The geography of happiness: Connecting Twitter sentiment and expression, demographics, and objective characteristics of place. *PLoS ONE, 8,* e64417.

Nagel, A. C., Tsou, M.-H., Spitzberg, B. H., et al. (2013). The complex relationship of Realspace events and messages in cyberspace: Case study of influenza and pertussis using tweets. *Journal of Medical Internet Research, 15,* e237.

Nara, A., Allen, C., & Izumi, K. (2017). Surgical phase recognition using movement data from video imagery and location sensor data. In D. A. Griffith, Y. Chun, & D. J. Dean (Eds.), *Advances in Geocomputation* (pp. 229–237). Berlin: Springer International Publishing.

Niedzielski, M. A., O'Kelly, M. E., & Boschmann, E. E. (2015). Synthesizing spatial interaction data for social science research: Validation and an investigation of spatial mismatch in Wichita, Kansas. *Computers, Environment and Urban Systems, 54,* 204–218.

Shaw, S.-L., Tsou, M.-H., & Ye, X. (2016). Editorial: Human dynamics in the mobile and big data era. *International Journal of Geographical Information Science, 30,* 1687–1693.

Shaw, S.-L., & Yu, H. (2009). A GIS-based time-geographic approach of studying individual activities and interactions in a hybrid physical–virtual space. *Journal of Transport Geography, 17,* 141–149.

Sui, D., & Goodchild, M. (2011). The convergence of GIS and social media: Challenges for GIScience. *International Journal of Geographical Information Science, 25,* 1737–1748.

Torrens, P. M. (2015). Intertwining agents and environments. *Environmental Earth Sciences, 74,* 7117–7131.

Torrens, P. M. (2016). Computational streetscapes. *Computation, 4,* 37.

Torrens, P. M., Nara, A., Li, X., Zhu, H., Griffin, W. A., & Brown, S. B. (2012). An extensible simulation environment and movement metrics for testing walking behavior in agent-based models. *Computers, Environment and Urban Systems, 36,* 1–17.

Tsou, M.-H. (2015). Research challenges and opportunities in mapping social media and big data. *Cartography and Geographic Information Science, 42,* 70–74.

Tsou, M.-H., Yang, J.-A., Lusher, D., Han, S., Spitzberg, B., Gawron, J. M., et al. (2013). Mapping social activities and concepts with social media (Twitter) and web search engines (Yahoo and Bing): A case study in 2012 US Presidential Election. *Cartography and Geographic Information Science, 40,* 337–348.

Wang, Z., Ye, X., & Tsou, M.-H. (2016). Spatial, temporal, and content analysis of Twitter for wildfire hazards. *Natural Hazards, 83,* 523–540.

Yang, J.-A., Tsou, M.-H., Jung, C.-T., Allen, C., Spitzberg, B. H., Gawron, J. M., et al. (2016). Social media analytics and research testbed (SMART): Exploring spatiotemporal patterns of human dynamics with geo-targeted social media messages. *Big Data & Society, 3,* 2053951716652914.

Yuan, M., & Nara, A. (2015). Space-time analytics of tracks for the understanding of patterns of life. *Space-Time Integration Geography and GIScience,* 373–398.

Yuan, M., Nara, A., & Bothwell, J. (2014). Space–time representation and analytics. *Annals of GIS, 20,* 1–9.

Zhao, Z., Shaw, S.-L., Xu, Y., Lu, F., Chen, J., & Yin, L. (2016). Understanding the bias of call detail records in human mobility research. *International Journal of Geographical Information Science, 30,* 1738–1762.

Zhao, B., & Sui, D. (2017). True lies in geospatial big data: Detecting location spoofing in social media. *Annals of GIS,* 1–14.

Author Biographies

Atsushi Nara (Ph.D., Arizona State University), Assistant Professor, Department of Geography, San Diego State University; Associate Director, Center for Human Dynamics in the Mobile Age (HDMA). His research interests are in spatiotemporal data analytics, modeling behavioral geography and complex urban and social systems, and geospatial computation. His recent research projects studies human movement behaviors including crowds and pedestrian dynamics under evacuation, massive GPS tracks of offenders, surgical staff in highly complex hospital environments, and residential relocation and gentrification dynamics by applying spatiotemporal data analytic algorithms, agent-based models, and web-based GIS.

Ming-Hsiang Tsou (Ph.D., University of Colorado, Boulder), Professor, Department of Geography, San Diego State University; Founding Director, Center for Human Dynamics in the Mobile Age (HDMA) (http://humandynamics.sdsu.edu). His research interests are in Big Data, Human Dynamics, Social Media, Visualization, and Internet GIS. He is co-author of Internet GIS, a scholarly book published in 2003 by Wiley and served on the editorial boards of the Annals of GIS (2008-), Cartography and GIScience (2013-) and the Professional Geographers (2011-). His recent NSF research projects focus on health disparities, disaster responses, and spatial social networks.

Jiue-An Yang (Ph.D., San Diego State University), Postdoctoral Research Scientist, The Qualcomm Institute and The California Institute for Telecommunications and Information Technology (CALIT2) at University of California, San Diego. His current research interests include geospatial artificial intelligence, machine learning for health intervention, and Big geo-data analytics. In the domain of human dynamics, he is specifically interested in methods for quantifying environmental exposure for individuals.

Cheng-Chia Huang (M.S., San Diego State University), Graduate Research Associate, Department of Geography, San Diego State University. Her research interests are in geocomputation, urban dynamics, and social media.

Chapter 13
Outlook and Next Steps: From Human Dynamics to Smart and Connected Communities

Daniel Sui and Shih-Lung Shaw

13.1 Summary/Recap

Although framed from a predominantly GIS and geographic perspective, the previous chapters in this volume represent some major advances in research related to human dynamics in recent years. The field seems to have been continuing expanding with growing interdisciplinary interests (Orsucci 2015; Wang et al. 2016; Zha et al. 2016; Shaw et al. 2016). In sharp contrast to Forrester's (1961, 1969, 1971) top-down and simulation-driven approach to studying industrial, urban, and world dynamics via systematic analysis almost half a century ago, the chapters in this volume and some of the latest papers on the topic (Wang et al. 2017; Yan et al. 2017) have not only signaled a recent surge in interdisciplinary study of human dynamics but also has witnessed a profound paradigm shift as defined by the following trends, which are worth recapping:

1. Instead of relying on simulation, human dynamics research is heavily data-driven. Almost all but one chapter use data harvested from various social media platforms (Gautam et al.; Xu). In addition, cell phone data (Xu et al.), GPS data from MicroSoft's Geolife (Miller and Hoover); SMS & CDR data are also used in studying human dynamics (Nara et al.)
2. Instead of aggregated analysis at certain geographic level (e.g. census tracts, TAZ units etc.), recent human dynamic research has much improved spatial and temporal granularity, mostly focusing on human at the individual level, potentially covering both indoor and outdoor mobility and network activities (Gao et al.; Xu; Wen et al.).

D. Sui (✉)
Department of Geography, Ohio State University, Columbus, OH, USA
e-mail: sui.10@osu.edu

S.-L. Shaw
Department of Geography, University of Tennessee, Knoxville, TN, USA

© Springer International Publishing AG, part of Springer Nature 2018
S.-L. Shaw and D. Sui (eds.), *Human Dynamics Research in Smart and Connected Communities*, Human Dynamics in Smart Cities,
https://doi.org/10.1007/978-3-319-73247-3_13

3. Moving away from merely empirical analysis, recent human dynamics research also tend to be more theoretically informed and guided. There seems to be a keen concern about the semantics and ontology (Koylu et al.), a shift from space to place and time (Cheng and Shen), and attempts have been made to link human dynamics research to a cybernetic framework (Xu).

4. As exemplified in the previous chapters, human dynamics research increasingly try to capture the link/connection between physical and virtual space, which is consistent with recent work on the O2O (on-line–offline) interaction and cyber-physical system (CPS) (Gao et al.; Xu et al.; Wen et al.)

5. Methodologically, we also witness the new advances in text mining/semantics (Nara et al.)—SVM/LDA, trajectory analysis, machine learning and data mining, time geography (Xu et al.), the resource description framework (RDF) (Fan and Stewart), and the new development of open-source tools (Ye et al.).

13.2 From Human Dynamics to Smart and Connected Communities

As we discussed in the introductory chapter, the advances in a plethora of technologies in information and communication (ICT), GIS/GPS, and sensor technologies etc. have transformed human communications in dramatic and profound ways. The new human dynamics as manifested at different scales are all directly or indirectly results of these technological advances. More than ever, humans are interconnected not only in the cities or communities they physically live but also increasingly linked together in virtual communities, or communities without propinquity defined by the space of information flows at the speed of light. Smart technologies have not only enabled this new round of research on human dynamics but also contributed to the birth and expansion of the so-called smart city and connected communities. The human dynamics constitute and are also constituted by emerging smart and connected communities in both physical and virtual spaces. Although implicitly, chapters in this volume (except Chap. 12) are also contributing broadly to our understanding of the dynamics of smart and connected communities.

Cities and communities in the U.S. and around the world are entering a new era of transformational changes, in which their inhabitants and the surrounding built and natural environments are increasingly connected by smart technologies, leading to new opportunities for innovation, improved services, and enhanced quality of life (Townsend 2013; Ratti 2015; Herzberg 2017). We believe that research on human dynamics is intimately related to the study of smart and connected communities. Studies on smart and connected communities should start with new understanding of human dynamics whereas new findings in human dynamics should be scalable and helpful for us to better understand the dynamics of smart and connected communities.

In many ways similar to the concept of human dynamics, the concept of smart and connected communities is also elusive. Rapid and pervasive technological changes are transforming urban life and urban governance around the world. The active engagement of city governments around the world with these changes is reflected by the fact that by the end of 2013 there were more than 143 smart city projects around the world and the number of smart city projects have been continuing to grow (Lee et al. 2014). Albino et al. (2015) noted that smart city is a "fuzzy concept (...) used in ways that [are] not always consistent. There is neither a single template of framing a smart city, nor a one-size-fits-all definition of it (page 4)". However, it is abundantly clear that smart city or smart and connected communities always entail the dual aspects—the smart technologies (i.e. the technological infrastructure) and the new governance and practices enabled by these technologies (i.e. the human side) (Kitchin 2014).

Until recently, research on smart and connected communities has been shaped and driven predominantly by the engineering and technology disciplines. The emphasis on the smart infrastructure has renewed the belief that smart governance of the 'data-polis' might translate in increased possibilities to steer our cities towards a more efficient operation. But any student of the history should recognize that no technological innovations have made the desirable social progresses without a concomitant change in human behavior and decision making process. In recent years there is an increasing interest for the smart city from the social sciences. But clearly, much remains to be done, especially coupling/linking the human dynamics study with the research of smart and connected communities. Without linking it more closely to studies of human dynamics, smart cities may risk excluding groups without capacities from participating in an increasingly technological world. As being demonstrated by the smart city literature so far, with its focus on algorithms and calculable data, and related processes of depoliticization, responsibilities become more opaque in the smart and connected communities. At the same time, possibilities for participatory governance evaporate if our focus is on the technical rather than the human side of the smart city. Apparently, these sort of issues will have a huge impact on urban life in the decades to come, we have only started to explore these socio-political issues (Leszczynski 2016). On top of this, much of the commentary and analysis is rooted in US and European experiences, yet arguably the on-the-ground transformations in terms of hardware and software are most evident in the fast paced urbanization of Asia and other parts of the developing world. One emerging mandate is to put the citizen and citizen participation at the front and center of smart city research. By integrating studies of human dynamics with research on smart and connected communities, we can surely make human dynamics research more scalable and also fulfill the mandate of human-centered study on smart and connected communities.

13.3   Outlook and Next Steps

The convergence of GIS, human dynamics, and smart and connected research is accelerating (Roche 2016). Moving forward, as we close the research gap between human dynamics and smart city, we venture to outline the following thoughts as discussion launching pad for the next phase of research on human dynamics and smart and connected communities, especially from a GIS and geographical perspective.

13.3.1   Integrating Organic with Designed Data: Moving Beyond the Big Data Hype

Big data has been trendy during the past 10 years in interdisciplinary research. To a large extent, recent advances in human dynamics research can be attributed to the big data deluge, as demonstrated in the previous chapters in this volume. However, we believe big data is an overly hyped term and it has lost its precise scientific meaning as what is big or small is all relative to our computing capabilities. Moving forward, a more useful distinction for human dynamics research is between designed data ('made' data) and organic data ('found' data). Designed data is the data collected through surveys, censuses, and administrative forms, while organic data refers to the self-generated 'big data' (Grove 2011). Designed and organic data are different in their stimulus to data production. Designed data are generated according to some specific, pre-defined data stimulus (the question), and the data will be applied to solve the stimulus question (Grove 2011). As for the organic data, however, data stimulus does not have to be determined before data production. As a result, for most of the time, the source data was generated to support its original application, but might also be valuable to another application as data are limited or not easy to obtain in this additional application field. However, as the original purpose of gathering the source data is not for the additional application, the data might not be suitable to be used directly. There are cases of this gap between the generation of data and the utility of data. For example, taxi trajectory data are primarily generated for security management, but could be used to identify passengers' travel behavior and traffic conditions; cell-phone data collected for fee management can also be used to estimate population distribution.

In our current rush to use big/organic data, we should not lose sight on the value and utility of small/designed data as they reveal different aspects of human dynamics. Major breakthroughs in human dynamics research will rely on the further integration of both organic and designed data.

13.3.2   Balancing the Positive and Normative Perspectives

Similar to the unreasonable effectiveness of mathematics of the early 20th century (Wigner 1960), we are witnessing the unreasonable effectiveness of data (Halvy et al. 2009) in the early 21st century, which has prompted calls for the end of theory in empirical research (Anderson 2008). Far from the truth, we need to have better, more rigorous theoretical framework to guide our empirical inquiries more than ever. Abundance of data is helpful, but data rarely speak for themselves unless and until they are put in proper theoretical and conceptual contexts. For human dynamics research, a balance between positive and normative inquiries is warranted.

In his recent work The New Science of Cities, Batty (2013) discussed the positive (focusing on "what is") dimension of the city science and the normative (focusing on "what should be") aspect of urban research. In the field of urban studies, the positive approach, or empirical approach, is related to urban modeling and simulation, while the normative approach, or idealistic approach, is more in line with the tradition of urban design. According to Batty (2013), these two dimensions of city sciences are not mutually exclusive, and their tools and methods could be effectively combined to examine urban issues.

For research related to human dynamics, the field is predominantly positive so far. Driven by scientific methods, human dynamics research aims to understand the emerging regularities of human mobility and activities. While the research along the positive tradition is crucial, it is also important to articulate a normative vision on what an optimal/ideal human dynamics should be and what changes we can and should do to facilitate the optimal human dynamics. In other words, we need to move human dynamics research from being descriptive (what is) to being more prescriptive (what should be). The positive and normative research of human dynamics are not mutually exclusive. Rather than simply being the passive users of technology, which is very bias-prone (The Guardian 2016), we can play an active role in designing the new system to incorporate insights gained from the positive research such as the recent work to remove gender biases from algorithms (Zou 2016). Findings from positive research regarding human mobility and behavior can inform normative research such as geodesign. One of Jay Forrester's (1998) last papers actually focuses on designing the future and how to make social system more sustainable. Given the accelerated trend towards ubiquitous computing, we believe that it should be fruitful to explore how to use geodesign to create the optimal human dynamics. Furthermore, we also need to creatively apply the insights gained from human dynamics research to better design our smart and connected communities.

13.3.3 Crossing the Quantitative Versus Qualitative Chasm: Towards a Mixed Method Approach

Methodologically, studies in human dynamics as of today are still driven predominantly by quantitative approaches, but the complexity of human dynamics calls for methodological hybridity to cross the quantitative versus qualitative chasm (Sui and DeLyser 2012). We believe a lot more productive research can be done through a mixed method approach by linking the spatial-analytical with the social-critical approaches.

Furthermore, we strongly believe that the emerging data-driven storytelling (Genauer 2016) should be an integral part of the tool box for human dynamics research. In fact, research storytelling is a much broader interdisciplinary trend that cuts across physical and social sciences as well as humanities, reflecting a renewed focus on the political possibilities afforded by storytelling (Sui 2015). Echoing this general trend of digital storytelling and data curation, one creative way to practice the so called qualitative GIS in recent years is through map stories or location-based storytelling more generally. Such map-based geospatial storytelling has empowered contemporary map makers to describe reality in a way unattainable by traditional quantitative approaches alone (http://mapstory.org; http://storymaps.esri.com/home). Story maps often integrate text, multimedia, and interactive functions to inform, educate, entertain, and inspire people. Story maps can empower a global community to organize knowledge about the world spatially and temporally. With platforms such as mapstory.org or storymaps.esri.com, people of all kinds can turn into story tellers who are capable of creating, storing, and collaborating on map stories and improve on understanding of global dynamics. Incorporating four Cs (countries, cities, companies, and communities) as integral components of geostrategy, GIS-based story-telling is helping people in all walks of life to better understand the world. With more and more platforms for location-based story-telling coming into existence, such as echolocation and storieseverywhere.org, we can tell a better story about the new human dynamics and smart and connected communities.

13.3.4 Striving for Efficiency, Equity, and Sustainability

As far as policy options are concerned, human dynamics research is disproportionately towards making the urban system more efficient, especially in the context of human mobility in our increasingly crowded cities. While these works on efficiency are commendable, future research should also give attention to equity and sustainability. As more and more smart technologies are replacing humans in the workplace, there will be new winners and losers. How can we use the smart technologies to create a more just and equitable society and to narrow the gap between the rich and poor? Furthermore, it's also quite murky and uncertain at this point at the environmental impacts of these smart technologies? Are they going to consume less materials and energy, thus more environmentally friendly or just the opposite? Recent studies on the emerging sharing economy show mixed results. For example, the growing popularity of Uber is not only further exacerbating the haves and have-nots but also caused extended hours of traffic congestion and further contribute to the deteriorating air quality. Apparently, there are challenging issues for human dynamics researchers to tackle in the coming years.

In response to these challenges, research on human dynamics should engage critically with the smart cities agenda that is in relation to the technological capabilities for more active urban citizenship and for more inclusive and responsive urban policy making. Are we allowing citizens to be participatory in the process of city design as well as smart city network? At the same time, it will explore the potential downside in relation to more centralized and less transparent decision making systems, 'automatized' and monetized control and distribution systems, more extensive surveillance and digital exclusion. Regardless, a more holistic policy purview to include concerns for efficiency, equity, and sustainability should serve us well both in the short as well as the long term.

13.3.5 Towards Robust and Reliable Research in Human Dynamics and Smart City

To make human dynamics and smart city research and practice gain the respect, credibility, and acceptance by the broader research community and policy makers, we strongly feel the next step for the human dynamics research is to make it more robust and reliable. This means that we need to intensify our efforts to make our research (to the best of our ability) to pass the acid test of reproducibility, replicability, and generalizability. This is the only way we make human dynamics more trust worthy and respected (Donoho 2014). According to a document released by an NSF interdisciplinary panel (NSF 2015a), robust and reliable scientific research must be reproducible, replicable, and generalizable:

- Reproducible: researchers can reproduce the results of a prior study using the same materials and procedures as used by the original researcher;
- Replicable: researchers can duplicate the results of a prior study if the same procedures are followed but new data are collected;
- Generalizable: researchers can empirically demonstrate that the results of a study are applicable in other contexts or populations.

Compared to other fields in physical sciences, engineering, and social sciences (King 2003; Chang and Li 2015; Camerer et al. 2016; Stodden et al. 2016), human dynamics research lags behind in our efforts to promote robust and reliable scientific practices. We still know little whether the voluminous studies published so far in human dynamics and smart city research based on organic data are reproducible, replicable, and generalizable. We need to conduct more replication and generalization studies of human dynamics research findings, and the practice of the emerging open science paradigm is apparently the first step towards the goals of robust and reliable science. The U.S. NSF has already mandated all funded projects to improve their public access plan (NSF 2015b). Refereed journals can also do more to promote robust and reliable sciences by requiring authors make their protocols and both organic and designed data available as a pre-condition for acceptance of publication, which can facilitate cross-validation between the findings from organic and designed data. Academic journals can further encourage reproduction and replication efforts by publishing negative results. Academic institutions can do their fair share by changing the incentive for success in science so that data-sharing and replication studies are rewarded as an integral part of the scientific research. However, we also should keep in mind that reproducible and replicable findings do not necessarily guarantee that the findings reflect the truth. It is relatively easy to replicate and reproduce similar results based on the same or similar open data and procedures published in a paper. However, the data could be biased or inappropriate methods could be used that lead to misleading results (Zhao et al. 2016). In the meantime, there often exist some unique characteristics of human dynamics beyond the generalizable results when similar data from different cities are compared. These remain as critical challenges to human dynamics and smart city researchers to develop robust and reliable research.

13.4   Conclusion

In the geographic literature, Finch (1939) was one of the earliest pioneers envisioned the study of human dynamics in geographic science. Forrester's work focused on the system dynamics in the 1960s and 1970s at the aggregate level, but dipped into a low point in the 1980s and 1990s. However during the past twenty years, interests in dynamics have resurged with the data avalanche and new level of theoretical sophistication (Barabasi 2005). Now the growing interests in smart and connected communities are signaling that study of human dynamics should be further integrated and embedded in a broader geographical setting. We have indeed come to a full circle. The convergence of interdisciplinary research mandates that we need to make our future research scalable from human dynamics to smart and connected communities.

The U.S. National Science Foundation created a cross-directorate Smart and Connected Communities (S&CC) program in 2016, whose goal is "to support strongly interdisciplinary, integrative research and research capacity-building activities that will improve understanding of smart and connected communities and lead to discoveries that enable sustainable change to enhance community functioning." (https://www.nsf.gov/pubs/2016/nsf16610/nsf16610.htm) It clearly indicates the complex and interdisciplinary nature of the challenges. The U.S. National Science

Foundation recently also proposes 10 big ideas for future NSF investments (https://www.nsf.gov/about/congress/reports/nsf_big_ideas.pdf). Among them, the "Work at the Human-Technology Frontier: Shaping the Future" which is about the changing ways of producing goods, providing services, and collaborating with colleagues in relation to changing education to the preparation of future work force, the "Harnessing Data for 21st Century Science and Engineering" which supports basic research of enabling data-driven knowledge discovery and development of national research data infrastructure for the deluge of data, and "Growing Convergence Research at NSF" which blends scientific disciplines and fosters robust collaborations to address complex research problems are closely related to the theme of this edited volume. These new national initiatives will further propel research on human dynamics and smart city to a new level of excellence and we can't wait to share with you the exciting research findings in human dynamics and smart and connected communities in the subsequent volumes of this book series.

References

Albino, V., Berardi, U., & Dangelico, R. M. (2015). Smart cities: Definitions, dimensions, performance and initiatives. *Journal of Urban Technology, 22*(1), 3–21.

Anderson, C. (2008). The end of theory: The data deluge has made the scientific method obsolete. Available on-line at: https://www.wired.com/2008/06/pb-theory/. Last accessed 15 July 2017.

Barabási, A.-L. (2005). The origin of bursts and heavy tails in human dynamics. *Nature, 435* (7039), 207–211. https://doi.org/10.1038/nature03459.

Batty, M. (2013). *The new science of cities*. Cambridge, MA.: MIT Press.

Camerer, C. F., Dreber, A., Forsell, E., Ho, Teck Hua, Huber, J., Johannesson, M., et al. (2016). Evaluating replicability of laboratory experiments in economics. *Science, 351,* 1433–1436.

Chang, A. C., & Li, P. (2015). *Is economics research replicable?* Sixty Published Papers from Thirteen Journals Say "Usually Not", Finance and Economics Discussion Series 2015-083. Washington: Board of Governors of the Federal Reserve System, http://dx.doi.org/10.17016/FEDS.2015.083.

Donoho, D. (2014). How much of our published research can we believe? Systematic failure, their causes, a solution. Available on-line at: https://simonsfoundation.s3.amazonaws.com/share/mps/conferences/Symposium_on_Evidence_in_the_Natural_Sciences/Donoho_slides.pdf. Last accessed on 13 December 2017.

Finch, V. C. (1939). Geographical science and social philosophy. *Annals of the Association of American Geographers, 29*(1), 1–28. https://doi.org/10.1080/00045603909357185.

Forrester, J. W. (1961). *Industrial dynamics*. Cambridge, MA.: MIT Press.

Forrester, J. W. (1969). *Urban dynamics*. Cambridge, MA.: MIT Press.

Forrester, J. W. (1971). *Industrial dynamics*. Cambridge, MA.: Wright-Allen Press.

Forrester, J. W. (1998). Designing the future. Available on-line, http://static.clexchange.org/ftp/documents/whyk12sd/Y_1999-03DesigningTheFuture.pdf. Accessed on 15 July 2017.

Genauer, R. (2016). The story of data driven storytelling. Available on-line at: https://medium.com/towards-data-science/the-story-of-data-driven-storytelling-65a02aac9d4. Accessed on 20 July 2017.

Grove, R. (2011). "Designed data" and "organic data". Available on-line at https://www.census.gov/newsroom/blogs/director/2011/05/designed-data-and-organic-data.html. Last accessed 23 January 2018.

Halevy, A., Norvig, P., & Pereira, F. (2009). The unreasonable effectiveness of data. *IEEE Intelligent Systems, 9*, 1541–1672. Available on-line at: http://www.summarization.com/~radev/767w10/papers/Week06/TextRepresentation/Halevy.pdf. Access 20 July 2017.

Herzberg, C. (2017). *Smart cities, digital nations: Building smart cities in emerging countries and beyond*. Petaluma, CA.: Roundtree Press.

King, G. (2003, February). The future of replication. *International Studies Perspectives, 4*, 443–499. Copy at http://j.mp/kFJH2q.

Kitchin, R. (2014). The real-time city? Big data and smart urbanism. *GeoJournal, 79*, 1–14.

Lee, J. H., Hancock, M. G., & Hu, M. (2014). Towards an effective framework for building smart cities: Lessons from Seoul and San Fransisco. *Technological Forecasting and Social Change*.

Leszczynski, A. (2016). Speculative futures: Cities, data, and governance, beyond smart urbanism. *Environment and Planning A, 48*(9), 1691–1708.

NSF. (2015a). Social, behavioral, and economic sciences perspectives on robust and reliable science, report of the subcommittee on replicability in science. Advisory Committee to the NSF Directorate for Social, Behavioral, and Economic Sciences—May 2015. https://www.nsf.gov/sbe/AC_Materials/SBE_Robust_and_Reliable_Research_Report.pdf.

NSF. (2015b). Today's data, tomorrow's discoveries: Increasing access to the results of research funded by NSF, released March, 2015.

Orsucci, F. (Ed.). (2015). *Human dynamics: A complexity science open handbook*. London, UK: Nova Science Pub Inc.

Ratti, C. (2015). *The city of tomorrow: Sensors, networks, hackers, and the future of urban life*. New Heaven, CT: Yale University Press.

Roche, S. (2016). Geographic information science II: Less space, more places in smart cities. *Progress in Human Geography, 40*, 565–573.

Shaw, S. L., Tsou, Ming-Hsiang, & Ye, X. (2016). Human dynamics in the mobile and big data era. *International Journal of Geographical Information Science, 30*(9), 1687–1693. https://doi.org/10.1080/13658816.2016.1164317.

Stodden, V., McNutt, M., Bailey, D. H., Deelman, E., Gil, Y., Hanson, B., et al. (2016, December 9). "Enhancing reproducibility for computational methods" with co-authors. *Science, 354*, 1240–1241 (6317).

Sui, D. Z. (2015). Emerging GIS themes and the six senses of the new mind: Is GIS becoming a liberation technology? *Annals of GIS, 21*(1), 1–13.

Sui, D. Z., & DeLyser, D. (2012). Crossing the qualitative-quantitative chasm I: Hybrid geographies, the spatial turn, and volunteered geographic information (VGI). *Progress in Human Geography, 36*(1), 111–124.

The Guardian. (2016). How Google's search algorithm spreads false information with a rightwing bias. Available on-line at https://www.theguardian.com/technology/2016/dec/16/google-autocomplete-rightwing-bias-algorithm-political-propaganda. Last accessed 7 August 2017.

Townsend, A. (2013). *Smart cities: Big data, civic hackers, and the quest for a New Utopia*. New York City: W.W Norton & Co.

Wang, C., Guan, X., Qin, T., & Yang, T. (2016). Modeling heterogeneous and correlated human dynamics of online activities with double Pareto distributions. *Information Sciences, 330*, 186–198.

Wang, S., Feng, X., Wu, Y., & Xiao, J. (2017). Double dynamic scaling in human communication dynamics. *Physica A: Statistical Mechanics and its Applications, 473*, 313–318.

Wigner, E. (1960). The unreasonable effectiveness of mathematics in the natural sciences. *Communications on Pure and Applied Mathematics, 13*(1), 1–14.

Yan, D., Wei, Z., Han, X., & Wang, B. (2017). Empirical analysis on the human dynamics of blogging behavior on GitHub. *Physica A: Statistical Mechanics and its Applications, 465*, 775–781.

Zha, Y., Zhou, T., & Zhou, C. (2016). Unfolding large-scale online collaborative human dynamics. *Proceedings of the National Academy of Sciences (PNAS), 113*(51), 14627–14632.

Zhao, Z., Shaw, S.-L., Xu, Y., Lu, F., Chen, J., & Yin, L. (2016). Understanding the bias of call detail records in human mobility research. *International Journal of Geographical Information Science, 30*(9), 1738–1762.

Zou, J. (2016). Removing gender bias from algorithms. Available on-line at: http://theconversation.com/removing-gender-bias-from-algorithms-64721. Last accessed 7 August 2017.

Author Biographies

Daniel Sui is Arts and Sciences Distinguished Professor and Professor of Geography, Public Affairs, Public Health, and Urban/Regional Planning at the Ohio State University (OSU). Since July 2016, he has been on an IPA assignment to serve as the Division Director for Social and Economic Sciences (SES) at the U.S. National Science Foundation. Prior his current appointment, Daniel Sui served as Chair of Geography (2011–2015) and as Director of the Center for Urban & Regional Analysis (CURA) (2009–2012) at OSU. He holds a B.S. (1986) and M.S. (1989) from Peking University and Ph.D. from University of Georgia (1993). His current research interests cover issues related to robust and reliable practices in spatially integrated social sciences and humanities, human dynamics in smart and connected cities, and location-based social media. Daniel Sui was a 2009 Guggenheim Fellow, 2014 Google Faculty Fellow, and 2014 recipient of the distinguished scholar award from the Association of American Geographers. Sui was also the 2015 Public Policy Scholar in residence at the Woodrow Wilson International Center for Scholars. More information about Sui's current research can be found at: https://www.researchgate.net/profile/Daniel_Sui.

Shih-Lung Shaw is Alvin and Sally Beaman Professor and Arts and Sciences Excellence Professor of Geography at the University of Tennessee, Knoxville. He also serves as the Interim Associate Provost for international education at the University of Tennessee, Knoxville. He received his B.S. degree from the National Taiwan University and his M.A. and Ph.D. degrees from the Ohio State University. His research interests cover geographic information science (GIScience), transportation geography, time geography, GIS for transportation (GIS-T), and space-time analytics of human dynamics. His recent research has focused on space-time analytics of human activities and interactions in a hybrid physical-virtual world based on various types of individual tracking data such as cell phone data, online social media data, vehicle tracking data, travel-activity survey data, and population migration data. His research has led to the development of a space-time GIS for representation, analysis, and visualization of individual activities and interactions in a hybrid physical-virtual space. Dr. Shaw is a Fellow of the American Association for Advancement of Science (AAAS). He also received the Edward L. Ullman Award for Outstanding Contributions to Transportation Geography from the Association of American Geographers (AAG) and served as the Head of the Department of Geography at the University of Tennessee, Knoxville.